Overlooked Genius

An Introduction to the Work of
Kurt Gödel

Qu Wenming

瞿文鳴

Cover design and artwork by Becky Jhu

Also by Qu Wenming

Special Relativity: A Graphic Approach

© 2022, Qu Wenming
All rights reserved

No, no, you're not thinking, you're just being logical.

Niels Bohr

CONTENTS

Preface — v

Chapter 1 Prelude — 1
briefly sets the stage for Gödel, and places his work in the context of the times.

Chapter 2 A Brief History — 3
describes the growing influence of the axiomatic method in the 19th century. We then look at the turmoil that erupted, and how various schools of thought dealt with it.

Chapter 3 Some Concepts in Logic — 12
presents the background in logic that is necessary to appreciate Gödel's work.

Chapter 4 The Completeness of First-order Logic — 25
begins our account of Gödel's many achievements. His Completeness and Compactness Theorems are central to Mathematical Logic.

Chapter 5 Arithmetic — 30
describes the axiomatisation of arithmetic and the Peano Axioms, necessary background for Gödel's Incompleteness Theorems.

Chapter 6 Formal Undecidability in Arithmetic — 37
examines what is probably Gödel's most famous theorem, the Incompleteness Theorem of Arithmetic. Like the Completeness and Compactness Theorems, it has far-reaching consequences.

Chapter 7 Consistency of Arithmetic — 57
examines the Second Incompleteness Theorem and its implications for the consistency of theories of arithmetic, a key question behind the crisis in Mathematics.

Chapter 8 Misconceptions About Incompleteness — 63
points out and clarifies some common misunderstandings surrounding the Incompleteness Theorems.

Chapter 9 Set Theory — 67
introduces some concepts in set theory, and poses two questions that remained unanswered until Gödel gave a partial solution.

Chapter 10 Cosmology — 82
describes Gödel's discovery that time travel is consistent with the laws of the General Theory of Relativity.

Chapter 11 Odds and Ends — 88
describes some of Gödel's lesser-known achievements.

Appendix A The First-order Induction Axiom Schema — 93

Appendix B Zermelo-Fraenkel Set Theory — 94

Appendix C Cardinal Numbers — 98

INDEX — 101

PREFACE

Kurt Gödel (1906-78) was an Austrian mathematician/logician, and one of greatest geniuses of the 20th century. For originality and depth of insight, he was at least the equal of Albert Einstein, and his discoveries had as great an impact on the foundations of Mathematics as Einstein's had on the foundations of Physics. But in terms of public recognition, Gödel is totally eclipsed by his famous friend. *Overlooked Genius* is an attempt to redress this imbalance. You may well ask, "Why should I try to learn something about the discoveries of Gödel?" The answer is that the process may bring you great pleasure. Why study Einstein's theories? Why listen to Beethoven? Of course, not everyone finds pleasure in understanding why $E = mc^2$, or in being swept up by the Ninth Symphony, so this book is not for everyone. But if you enjoy intellectual challenges and examining the thoughts of a great mind, it might be for you.

Since Gödel's work is well-known to anyone who has studied Mathematical Logic, this book is addressed to those with no background in the subject. But we must start somewhere, so I assume that you are familiar with high school mathematics. From this starting point, I hope to explain the importance of Gödel's theorems, and to provide some insight into how he proved them. We are fortunate that the essence of some of his ingenious methods can be understood (up to a point) by non-specialists.

As Bertrand Russell said, a book cannot be both intelligible and correct. Bearing in mind the intended readership, I have leaned toward intelligibility. Stating every mathematical concept and result that is mentioned with full correctness would have meant the inclusion of many technical details that would have taken up much of your time and effort; the gain would not be worth the cost, and the profusion of trees may even obscure the forest. So be aware that the mathematics described in this book is usually not the whole story. I hope that the details I have omitted are not essential in reaching a satisfactory understanding of the material. Some topics of a more technical nature are covered in greater depth in the Appendices.

This book covers very few aspects of Gödel's life. If you are interested in Gödel the man and not just his mathematics, you should read the excellent biography *Logical Dilemmas* by John W. Dawson Jr., (A K Peters, Ltd., 1997). It is thoroughly researched and very well written.

Another aim in writing this book is to dispel the commonly held belief that Mathematics is a cut-and-dried subject, with no room for diverse opinions. This may be true of the material taught in high school, but at the deepest levels, there has always been debate about what has been acceptably proven, and what concepts and methods are admissible as being "mathematical". These considerations are especially important to those studying the foundations of Mathematics, and several examples of serious disagreements are mentioned in this book. In the 21st century, some old questions remain unsettled, and new ones have arisen concerning the role of computers.

瞿文鳴
Qu Wenming
quwenming20@gmail.com

CHAPTER 1
PRELUDE

1.1 A Tale of Two Disciplines

It was the best of times, it was the worst of times,...(No, you haven't opened the wrong book.)

The decades that bracketed the end of the 19th century was such a time for both Physics and Mathematics. Cracks had been discovered in the foundations that threatened both structures. For bright young physicists and mathematicians, it was the best of times, exciting times that would herald a revolution in their field. For the established old guard, it was the worst of times, watching in anguish as a lifetime of work came under threat from unforeseen developments.

Both disciplines had enjoyed a period of solid progress. Since the early 19th century, mathematicians had been re-shaping the subject as a confident new vision gained traction (more on this in the next chapter). In Physics, James Maxwell's equations of electromagnetism appeared to some to signal the end of theoretical Physics; all phenomena were now understood. But tremors were being felt around the two dormant volcanoes. When they erupted, both disciplines were in crisis.

Times of crisis bring forth new leaders. The renovation of Physics was led by Albert Einstein and Niels Bohr. They did much more than patch the cracks; they rebuilt the entire foundations, and the house of Physics would never look the same again. But while they both succeeded brilliantly, a large gap remains between the work that each of them did. But that is a story for another book.

In Mathematics, one prominent leader was David Hilbert, professor at the prestigious University of Göttingen. He drew up plans to shore up the foundations, and the brilliant young John von Neumann was one of several who took up the task. (He was also involved with Bohr's rebuilding job in Physics, a testament to the breadth and depth of his talents.)

1.2 A Tale of Three Geniuses

As von Neumann worked to repair the foundations according to Hilbert's plans, Kurt Gödel entered the picture. Although it was not his initial intention to do so, he showed that crucial parts of the plans could not be completed; the architect had been too ambitious. Gödel's discoveries had profound implications, but they were not immediately recognized. One of the first to understand them and to appreciate their significance was von Neumann. He enthusiastically publicised Gödel's achievements, something which Gödel, by nature, was not well-equipped to do.

Both men were highly prolific, von Neumann probably even more so than Gödel. The former not only worked with Hilbert, but made significant contributions to many fields: quantum mechanics (Bohr's domain), computer science, economics, and nuclear energy, to name just a few. Gödel remained mostly in the field of Mathematical Logic, but made important discoveries in many areas.

The Institute for Advanced Study (IAS) in Princeton, USA is a research centre that attracts the best and brightest scientists and mathematicians, encouraging them to do basic research for its own sake. According to Dawson (p. 201), when the IAS showed hesitancy in promoting Gödel to a professorship, von Neumann (himself a professor) said, *"How can any of us be called professor when Gödel is not?"*.

At the IAS, Gödel and Einstein became close friends. Einstein is reported to have said that he went to the office *"just to have the privilege of walking home with Kurt Gödel"*. How fascinating it would have been to hover around them as a fly (one that understands German) during their walks. Perhaps inspired by his friend, Gödel briefly returned to his first love, Physics, and found a radically new solution to Einstein's equations of General Relativity. Einstein's death came as a shock to Gödel. His fragile health and mental stability were dealt another blow by the premature death of von Neumann just two years later. Without the two friends whose calibre matched his own, Gödel nonetheless carried on his work, albeit with diminished energy, until his own death twenty years later.

In the next chapter, we look at the evolution of Mathematics in the 19th century, before inspecting the cracks that appeared in the foundations. We will also need to understand the plans that Hilbert drew up, and to be familiar with a few concepts in Mathematical Logic. Then we will be ready to look into the work of Kurt Gödel.

CHAPTER 2

A BRIEF HISTORY

2.1 Mathematics Unchained

During the 19th century, Mathematics was freed from the yoke of physical reality. The legitimacy of a mathematical statement had traditionally relied on how accurately it described physical reality; mathematical truth was not clearly distinguished from physical truth. The discovery of non-Euclidean geometry propelled the movement away from that viewpoint.

Euclidean geometry is founded on five axioms, and the fifth one states that given any line and any point not on the line, there is exactly one line through the point parallel to the line. All the axioms were considered to be "obviously true" in physical space. In the early 19th century, Carl Friedrich Gauss realized that a perfectly logical system could be obtained by replacing the fifth axiom with one asserting that more than one line could pass through the given point parallel to the given line. He had developed non-Euclidean geometry, but chose not to publish his discovery, fearing that it would stir up controversy, or even expose him to ridicule. A few years later, Nikolai Lobachevsky and János Bolyai independently came to the same idea, and each published his findings. (This is not the only time that a bold new idea occurred to several people almost simultaneously.) Acceptance from the mathematical community was neither swift nor universal, but the genie was out of the bottle. By the end of the century, there was general acceptance that a theory requires only axioms that are consistent, free from contradiction. Moreover, terms such as "point" and "line" should remain undefined because logical conclusions are drawn based on the structure of statements, not their meaning. Definitions like "a point is that which has no extension" are of no use in a proof, and could be misleading. After centuries of dependence on physical reality, Mathematics was free at last.

This liberation did not just happen overnight in the manner of a religious conversion, or through a declaration of independence. In the axiomatic method, proofs are driven by logic. But the logic that was inherited from ancient times was not as powerful as what mathematicians were using; logic needed to be strengthened before the axiomatic method could handle Mathematics. Around 1840, George Boole began the task of "mathematising" logic. Any situation involving an organized structure can be clarified and better understood by expressing the internal relationships in general and abstract terms, usually aided by the use of symbols. For example, algebra increases our understanding of the relationships between numbers, and geometry does the same for spatial relationships. Boole did the same for logic, creating a sort of algebra, known today as "Boolean algebra". The task of strengthening logic was completed in the mid-19th century by Gottlob Frege and Charles Peirce, among others.

While the axiomatic method coupled with the new logic granted mathematicians a fresh point of view, progress was being made on other fronts. In Mathematics, "analysis" refers to the study of real numbers. These are the numbers expressible in decimal format, using potentially infinitely many digits. Since the time of Newton and Leibniz, calculus (a part of analysis) had proved remarkably useful as a tool for physicists doing their calculations. But why the tool worked so well, or even worked at all, was unclear to mathematicians because calculus was not based on rigorously defined concepts. During the 19th century, this rather scandalous state of affairs was remedied by Augustin-Louis Cauchy and Karl Weierstrass, among others.

Arithmetic (or number theory) also embraced the new axiomatic spirit. The great theorems of Fermat, Euler, and Gauss had been proved without "official" axioms, and although no-one thought they were incorrect, Giuseppe Peano produced a set of axioms for arithmetic. Meanwhile, Frege was completing his *Basic Laws of Arithmetic*, a monumental work that was expected to provide a logical and rigorous basis for arithmetic.

By 1900, some were feeling confident and optimistic. Henri Poincaré went so far as to declare, *"One may say today that absolute rigour has been attained"*. But dark clouds had already gathered, and Hilbert saw them.

2.2 The Storm

The first signs that not all was well surfaced in the 1870s. Georg Cantor was working on the convergence of trigonometric series, a hot topic in analysis, and was led to consider infinitely large sets of real numbers. He defined what it meant for one infinite set to be larger than another, and proved that the set of natural numbers was just as large as the set of fractions, but that both were smaller than the set of real numbers. This led to a storm of protest. Considering infinite sets as completed objects that could be compared was not allowed! His work was rejected by many, even though none could find any errors in it. The hostility against his work even became personal, and Cantor suffered a nervous breakdown. But other mathematicians praised his work, and urged him to continue. When mathematicians cannot agree amongst themselves on what they should or should not be studying, there is a problem.

Then Cantor himself found another problem. He had shown with irrefutable logic (which we will see in Chapter 9) that any set is smaller than the set of all its subsets. Then what about the set of everything? It too must be smaller than the set of all its subsets, but the set of everything cannot be smaller than something. This contradiction is known as "Cantor's Paradox". The trouble is that the "set of everything" is a self-referential definition; a set is defined by its members, so as a member of itself, it is part of its own definition. Soon, all sorts of paradoxes based on self-referential constructions like this emerged. Bertrand Russell's barber declared that he shaved all those, and only those, who did not shave themselves. Russell asked if the barber shaved himself. If he did not, then according to his declaration, he did, which is a contradiction. But if he did shave himself, then again according to his own declaration, he didn't. Either way there is a contradiction. The barber's declaration is self-referential because he himself is one of "those" to whom he refers. We can dismiss Cantor's "set of everything" as something that cannot exist, and we can conclude that the barber's declaration must be false. But self-referential definitions can be found in accepted classical Mathematics. Are there objects that are defined in Mathematics that cannot exist? Are there theorems that are false declarations? For Hilbert and some others, this was a cause for concern.

Their concern was justified. In 1902, a self-referential contradiction was found where it was least expected. Frege's *Basic Laws of Arithmetic* was supposed to set arithmetic on a firm logical footing. But just before Volume II was to be published, Russell wrote to Frege informing him of a contradiction in the book, later known as "Russell's Paradox". Frege had allowed sets to be defined by properties; any property defines the set of all those sets that have this property. For example, the set of all those sets that have exactly 4 members includes the set of prime numbers less than 8, the set of Beatles, and many other sets. Russell considered the set of all those sets that have the property of not being members of themselves. Is this set a member of itself? If it is, then it fails to meet the condition for membership, and therefore it isn't. If it isn't, then it satisfies the condition for membership, and therefore it is. Either way, there is a contradiction. The barber had been hiding in Frege's book! (For the sake of historical accuracy, it must be noted here that Russell's Paradox was discovered several years before the Barber's Paradox.) The discovery of Frege's error came as a shock because he was such a highly regarded mathematician, yet his notion of set was vulnerable to contradictory self-referential definitions.

Mathematicians were now combing through existing proofs, looking for self-referential definitions and checking for errors. If Frege could make a mistake like this, anyone could. What they found was that in a number of accepted classical theorems, sets were assumed to exist based on rather dodgy grounds. In some cases, mathematicians (even Cantor himself) were unaware that they were treading on thin ice. We will look at this in later chapters, but for now, we just note that in some guises, the dodgy assumption appeared very plausible. Yet when handled with cunning, this plausible way of creating sets led to very counter-intuitive consequences. In 1904, Ernst Zermelo took the bull by the horns, and formalised this shaky way of producing sets by formulating the Axiom of Choice. In a replay of what befell Cantor thirty years earlier, a flood of protest descended upon Zermelo (but thankfully, not much personal abuse). This was not mathematics, cried the opponents. Yet the proofs of many theorems relied on the Axiom of Choice. For example, without the Axiom, it is not possible to prove that every vector space has a basis. Is the Axiom true or not? Might it be provable? Might it be false? The foundations of Mathematics appeared to be afflicted by inconsistency and uncertainty. In a few short years, the confidence of 1900 had vanished.

2.3 The Constructivist Reaction

When new ideas bring sudden changes to an established order, it tends to push back. This phenomenon has been seen in politics, religion, culture, and other spheres of human activity. Mathematics was no exception; the ground-breaking changes seen in the 19th century and the problems they engendered triggered a reaction.

Most of the objections centered around the idea of the infinite. The laws of logic or any properties of mathematical objects are based on human experience. But we encounter only what is finite, and have no direct experience with the infinite. (Let's keep religion out of this narrative.) (The last parenthetical remark is self-referential; it says something about a narrative while it is part of the same narrative.) Yet mathematicians have blithely (and blindly?) extended what they know about the finite to the infinite. For some, that is a leap too far.

At the beginning of the 20th century, this school of thought became known as *"constructivism"*. Leopold Kronecker, who objected vehemently to Cantor's work on infinite sets in the 1870s, was a constructivist before the term was coined. In particular, the constructivists objected to the practice in traditional Mathematics that deduced the existence of entities on the grounds that their non-existence was contradictory. They insisted that an entity is proved to exist only when a method of constructing it or finding it is described. The traditional argument that "an entity either exists or does not" was rejected by constructivists. In other words, they did not accept that for *every* statement **S**, either **S** or not-**S** must be true. This is a denial of the Law of the Excluded Middle, which had been used by mathematicians for centuries. There were several levels of constructivism, defined mostly by how thoroughly one banished involvement with the infinite. One prominent version, called *"intuitionism"*, was led by L.E.J. Brouwer and Arend Heyting. But using only constructivist logic, many theorems of traditional Mathematics are unprovable. In some versions, one could not even prove that every subset of a finite set is finite. The majority of mathematicians rejected constructivism on the grounds that it was too restrictive; it had thrown out the baby with the bathwater.

2.4 The Formalist Reaction

For Hilbert, the question of consistency was the main issue. In 1923, he laid out the formalist program, a way to re-establish confidence in the consistency of Mathematics. The key question was, are the various branches of Mathematics consistent? For example, can a contradiction be proved in arithmetic (i.e. number theory)? To answer such questions, he proposed to study the properties of proofs mathematically, as Euclid had done for geometry, and Boole had done for logic; each of them had "mathematised" a structured body of knowledge, creating a branch of Mathematics. Just as number theory is the study of numbers, Hilbert's proposed new branch of Mathematics is the study of proofs. It was thus called *"proof theory"*. Since proofs are the essence of Mathematics, proof theory was the "mathematisation" of Mathematics itself. Thus, another name for it was *"metamathematics"*, where the prefix "meta" means "about", or "concerning". If we consider Mathematics to be a snake that had swallowed arithmetic, geometry, algebra, analysis, logic, topology, set theory, and all the other domains of knowledge that make up its many branches, then Hilbert was asking the snake to bite on its own tail, and start swallowing itself. This is only a metaphor, but it illustrates the need to proceed carefully and avoid ending up in a logical quagmire of circular reasoning.

Hilbert's proposal involves two categories of proof and we need to separate them clearly. In the first category are the hundreds of existing proofs of Mathematics. The second category consists of the proofs in the new field of metamathematics. They prove things about the scope and limitations of proofs in the first category, or in other words, what can and cannot be proved in various branches of Mathematics. For example, we hope that a metamathematical theorem will prove that arithmetic cannot ever prove a contradiction, which would establish the consistency of arithmetic. Another metamathematical theorem might do the same for geometry. Although the question of consistency was the major concern, there were also other questions of interest to proof theorists. For example, is every true statement about numbers provable in arithmetic? This is not a question about numbers, so it is not a question for number theorists to consider. It is a metamathematical question.

Note that proof theorists are not interested in checking individual proofs in any other branch of Mathematics. For example, whether or not any particular proof in arithmetic is correct is a question for number theorists, not proof theorists. But whether or not a proof depends on the Axiom of Choice would be of interest to proof theorists.

Suppose some proof theorists are investigating proofs in arithmetic. They might begin by opening a textbook on number theory and looking at the proofs. The proofs they see are very informal; they typically contain a lot of verbal reasoning supported by a few equations, and there is heavy reliance on what the reader is expected to already know. It is highly unlikely that any of the proofs have any errors, but their unwieldy form makes it very difficult for the proof theorists to investigate the scope and limitations of proofs in arithmetic. How can one discover what can and cannot be proved by informal verbal reasoning?

To get around this difficulty, proof theorists took advantage of the fact that proofs are supposed to conform to a strictly defined format. Such proofs are called *"formal proofs"*, and we shall investigate them in the next chapter. For now, just note that formal proofs consist of statements that are assumed to be true (called *"axioms"*), and statements that are derived from preceding ones using *rules of inference*. The language used in a formal proof is also strictly defined. Every branch of Mathematics has its own language and axioms which determine what can be formally proved. When proof theorists wish to investigate the proofs of arithmetic, for example, they can examine the axioms of arithmetic and what the rules of inference can do with them. The rules are rigid and clearly defined, and it is absolutely clear what each rule can do. Whether or not a sequence of statements is a (formal) proof can be checked by a routine "mechanical" procedure that requires no thought or intuition, and is guaranteed to terminate; even a computer could do it. With formal proofs, there is none of the vagueness that is found in the verbal reasoning of informal proofs. Now proof theorists can systematically study what can and cannot be formally proved in arithmetic.

But proof theorists have now restricted themselves to considering only the mechanism behind formal proofs. Do their conclusions tell us anything about the scope and limitations of informal proofs, the ones that are found in the real world? The answer is "Yes!" because formal proofs and correct informal ones prove exactly the same theorems. Any correct informal proof can be translated to a formal one that proves the same theorem. (That may be taken as the definition of a "correct informal proof".) When proof theorists prove that some branch of Mathematics is consistent, what they have really proved is that no *formal* proof of a contradiction in that branch can exist. But that means there can never be a correct informal proof of a contradiction either. In the final analysis, every informal proof rests its case on a formal one, but this "certification" is seldom, if ever, demanded. Pedantic mathematicians who insist on asking "How do you know that?" at every stage of a verbally presented informal proof will be satisfied only when shown a formal proof of the same theorem (if they have not already been evicted from the lecture hall).

Formal proofs are never used in practice because they would be very tedious to write down, and what is worse, the reader would gain little insight into how the proof was discovered. Even proof theorists, whose main concern is what can and cannot be formally proved, hardly ever see a formal proof.

As he laid out the formalist program, Hilbert was aware of a difficulty. We are investigating the formal proofs in some branch of Mathematics because we hope to prove that this branch is consistent. If it were inconsistent, then it could prove nonsense, so if we have any doubts about its consistency, then we cannot fully trust its proofs. But our metamathematical conclusions are reached using proofs, the very instruments that we are investigating. What if proof theory itself is inconsistent? If we have any doubts about its consistency, we cannot trust its theorems. Are we, in effect, asking that a possibly corrupt politician be investigated by a colleague who is also under suspicion? Hilbert's solution was to insist that the proofs of proof theory must use very conservative methods, avoiding anything dubious like the Axiom of Choice, or the use of infinite sets. (Although Hilbert accepted Cantor's work on infinite sets, they were still regarded with a degree of suspicion.) These trustworthy methods of proof were called *"finitary"*, and they would certainly be trusted by most mathematicians, and even by some constructivists. The proofs of proof theory are informal, just like the proofs in other branches of Mathematics; informal proofs are no less reliable than formal ones, especially when only finitary methods of proof are used. What sets proof theory apart from other branches is the emphasis on finitary methods of proof. The person appointed to investigate our possibly corrupt politician must be widely recognized as being impeccably honest.

2.5 Set Theory and Logicism

We now take a look at the development of set theory. In the 1870s, Cantor was the first to study sets as mathematical objects and made the initial discoveries that founded set theory as a new branch of Mathematics. It was a "theory" in the informal sense of the word. Cantor worked with sets intuitively, and did not specify axioms for set theory. Despite receiving criticism from some quarters, his work gained acceptance with others. In particular, Frege and Richard Dedekind recognized the concept of a set as being a fundamental notion upon which Mathematics could be built. The idea of forming objects (called "sets") as collections of other objects was even more basic than the idea of numbers. Indeed, Dedekind was able to define natural numbers, rational numbers (fractions), real and complex numbers, all in terms of sets. This led him and Frege to take the position that Mathematics consisted of pure logic plus the notion of a set. No other "mathematical" concepts were needed, so if the notion of a set were regarded as a concept of logic, then mathematical truths are no more than logical truths. This school of thought became known as *"logicism"*, and Frege's *Basic Laws of Arithmetic* was written to support it, relying on an intuitive treatment of set theory.

But as we saw, Russell (ironically another logicist) found a fatal flaw (Russell's Paradox) in Frege's treatment of sets. Russell sought to rescue logicism from his own paradox, and together with Alfred North Whitehead, wrote *Principia Mathematica* (or *PM*), published in 1910. This enormous work was intended to support the central thesis of logicism that Mathematics is logic (set theory being a part of logic). The set theory in *PM* was an axiomatic theory that appeared to be consistent, but it was so cumbersome that it did not appear to be part of logic. Nonetheless, *PM* was a powerful system that gave a thorough exposition of the foundations of Mathematics.

In 1908, Zermelo produced a set of axioms for set theory. His work was later modified by Abraham Fraenkel and von Neumann, creating Zermelo-Fraenkel set theory (or **ZF**). It was simpler than the set theory of *PM*, and it appeared to be consistent. It was also powerful enough to incorporate essentially all of Mathematics, as Frege and Dedekind had done with informal (non-axiomatic) set theory.

By adding the Axiom of Choice to **ZF** as an extra axiom, we have the theory known as "**ZFC**". **ZFC** plays a role in making metamathematics a precise discipline. Today, it is widely accepted as the arbiter of open mathematical questions, a sort of "Supreme Court of Mathematics". Is there a function with this peculiar property? Is this dubious conjecture provable? Instead of relying on informal intuition (which led to the acrimonious debates of the 19th century), mathematicians now turn to **ZFC** for the answers. But this means that questions that are unanswerable by **ZFC** are mathematically unanswerable. To deal with such questions, one option is to strengthen **ZFC** by adding axioms that appear to be "true". But no consensus has emerged on what they might be.

2.6 Gödel

By 1923 when Hilbert laid out the formalist program, there were three well-defined viewpoints on the foundations of Mathematics: logicism, constructivism, and formalism. Their differences were deep and irreconcilable, but for most mathematicians, this was not an issue; they continued refining their concepts and proving theorems in their chosen fields, unaffected by the three competing "isms" at the root of their subject. Into this environment came Kurt Gödel, seventeen years old, philosophically inclined, and destined for a life in Mathematics. His interest soon turned to the unsettled questions on the foundations of Mathematics, and remained there for the rest of his life.

CHAPTER 3

SOME CONCEPTS IN LOGIC

3.1 Proofs and Logic

We first examine the structure of proofs in general, and the "old" logic. Then we will see how it was strengthened in the 19th century to meet the needs of Mathematics.

The structure of proofs in Philosophy and Mathematics is based on *axioms* and *rules of inference*. Axioms are statements that may be asserted without justification. A *proof* is a finite sequence of statements in which each line is either an axiom, or is derived from previous lines by a rule of inference. The first line in a proof must be an axiom because there are no previous lines on which to apply any rules. The rules of inference are set up to preserve truth based upon the *structure* of one or more previous statements in the proof. When applied to true statements, rules of inference always produce true statements. Therefore, *if the axioms are all true, then every line in a proof is true* because the first line is true (being an axiom), and any subsequent line is either an axiom or follows from previous lines by applying rules that preserve truth. If some axioms are false, then some lines in a proof may be false. The last line of a proof is called a *"theorem"*.

The following example of a proof will clarify what we mean by rules "preserve truth based upon...*structure*":

1.	Axiom 1.	If Jill is hungry then Bill is happy.	J→B
2.	Axiom 2.	Jill is hungry and sugar is sweet.	J&S
3.		Jill is hungry.	J
4.	Conclusion.	Bill is happy.	B

The above proof is stated in English, and in symbols. It begins with two axioms, and Line 3 is the first that uses a rule of inference. When **J&S** is true, then **J** must also be true based on the *structure* of **J&S**, and the *definition* of "and" (symbolized by &). We don't need to know the meaning of **J**, or when Jill had her last meal. This rule is called the "&-Elimination Rule" because it allows us to eliminate an &. It says that when you have a line in your proof that consists of an & between two statements, then the next line may be either one of the two statements. Line 4 is the result of applying the →-Elimination Rule to Lines 1 and 3. The structure of these two lines and the definition of → ensure that Line

4 is true if Lines 1 and 3 are. The symbols &, V, and → mean "and", "or", and "implies" (or "if..then..") respectively. Placing any of them between two statements gives us a new one. The symbol ¬ means "not", and it gives us a new statement from only one existing statement. These symbols are called "propositional connectives". Each of them is defined by the effect it has on the statement or statements to which it is applied. For example, by *definition*, $\alpha \to \beta$ is false when α is true and β is false, and it is true in all three other cases. Each connective has rules of inference which preserve truth based on the definition of that connective.

Logical axioms are statements that are always true by virtue of their structure. For example, $\alpha \lor \neg\alpha$ (which says α or not-α) is true for any statement α. So for any statement α, this is an axiom that may be used in any proof.

Proofs in Mathematics probably originated with the Greeks. For example, they looked at square numbers (those equal to some number squared) and noticed that 7^2 is almost double of 5^2. It was natural to ask if any square number is exactly double of another. After failing to find any, they would have suspected that such numbers do not exist. It is impossible to check every pair of square numbers, so the only way to be sure was through a proof. In Euclid's *Elements* (which appeared around 300 BCE), there is a proof of this assertion: no square number is the double of a square number. He used some obvious facts about numbers as axioms, and proceeded by rules of inference.

The system of proof that we have described, called "propositional logic", is not as powerful as the logic used by mathematicians. For example, propositional logic is not able to derive the conclusion below from the two axioms.

1. Axiom 1.	Daffy Duck is a bird.	**D**	**[3.1]**
2. Axiom 2.	All birds can fly.	**A**	**[3.2]**
....			
n. Conclusion.	Daffy Duck can fly.	**F**	**[3.3]**

Propositional connectives can treat these lines only as independent statements **D**, **A**, and **F**. The rules of inference can do nothing with them because the propositional connectives cannot "sink their teeth" into them. Logic needs to be strengthened, with a larger vocabulary and more rules of inference before we can formally prove the above deduction. To work with the above statements, the words "bird" in Line 1 and "birds" in Line 2 must be linked. We do this by defining a *predicate*. A predicate is a property that some things might possess, and others might not. We define *Bird*(__) to mean "__ is a bird". *Bird*(__) is a predicate that expresses the property of being a bird. By filling in the blank space, the predicate then says that the thing we named is a bird, which might be true, or false, or ambiguous. For example,

    Bird(Daffy Duck)   means Daffy Duck is a bird, which is true,
    Bird(Popeye)   means Popeye is a bird, which is false,
    Bird(It)   means It is a bird, which is ambiguous.

The word "It" is a pronoun and its meaning is variable. In Mathematics, we call such a word a *"variable"*, and we use a sans serif letter (e.g. x) instead of "It". There is usually some restriction on what objects a variable can mean, and the set of possible objects is called the *"universe"*.

In our present discussion, we take the universe to be the set of all things that ever lived, in fact or in fiction. We define the predicate *CanFly*(__) to mean "__ can fly". Now consider this statement:

    If *Bird*(x) then *CanFly*(x).

When written out fully in English, this says that

    If x is a bird then x can fly.   **[3.4]**

To most people, [3.4] colloquially means "All birds can fly", which is unambiguous. But strictly speaking, [3.4] is ambiguous because we have not specified what x is, and the meaning of [3.4] depends what we mean by x. We will be strict, and consider [3.4] to be ambiguous until we specify what x means. Choosing a meaning for x is done by replacing x with the name of some object in the universe. What [3.2] (All birds can fly) means is that we can replace x in [3.4] by the name of *any* object in our universe, and we will have a true statement if [3.2] is true. In other words, [3.2] says that

    For all x in our universe, if x is a bird then x can fly.   **[3.5]**

Expressing [3.5] symbolically, we have

    $\forall$x (*Bird*(x) $\to$ *CanFly*(x))   **[3.6]**

$\forall$ means "For all ...". It is called the *"universal quantifier"* and it must be followed by a variable. In [3.6], this variable x may not be replaced by the name of an object, so [3.6] is not ambiguous. There are rules of inference dictating how $\forall$ may be manipulated in a proof. The $\forall$-Elimination Rule says that if we have a line in our proof that begins with $\forall$, the next line in our proof may be the statement we get by removing both the leading $\forall$ and the adjacent variable, and then replacing all occurrences of that variable by the name of one object from our universe. For example, suppose

$\forall$y (y *likes* Daffy → Bugs *likes* y)

is a line already in our proof. Then the next line in our proof may be

Popeye *likes* Daffy → Bugs *likes* Popeye.

(We first removed $\forall$y, and then replaced every remaining y by Popeye.) The $\forall$-Elimination Rule captures the meaning of "For all ...".

We can now prove that Daffy Duck can fly, assuming Axioms 1 and 2:
1. Axiom 1. *Bird*(Daffy Duck).
2. Axiom 2. $\forall$x (*Bird*(x) → *CanFly*(x)).
3. *Bird*(Daffy Duck) → *CanFly*(Daffy Duck).
4. Conclusion. *CanFly*(Daffy Duck).

Applying the $\forall$-Elimination Rule to Line 2, and replacing x by Daffy Duck gives Line 3. Applying the →-Elimination Rule to Lines 1 and 3 gives Line 4.

In the above proof, Axiom 2 is false (kiwis can't fly), so we can't be sure that the Conclusion is true. In a proof, what we know for sure is that *if all the axioms are true, then the conclusion must be true.*

There is another quantifier called the *"existential quantifier"*. It is symbolized by $\exists$, and it means "There is something in our universe...". For example,

$\exists$x (*Bird*(x) & ¬*CanFly*(x)) [3.7]

says there is at least one object in the universe that is a bird that cannot fly. Suppose **A** is the name of some object in the universe. The $\exists$-Introduction Rule says that if *Bird*(**A**) & ¬*CanFly*(**A**) is a line already in the proof, then [3.7] may be the next line in the proof.

In a formal language, a *statement* is a predicate, or built up from other statements using quantifiers and propositional connectives. Here are three statements from arithmetic using the *"Less than"* predicate, denoted by "<". Our universe is $\mathbb{N}$, the set of natural numbers (0, 1, 2, 3, ...etc.).

x < 4 → 5 < y means if x is less than 4 then 5 is less than y,
$\forall$x (x < 4) means every number is less than 4,
$\forall$k $\exists$y (y < k) means for every number there is a smaller one.

An unquantified variable is called a *"free variable"*; it is free to assume different values, so statements with free variables are ambiguous. A statement in which every variable is quantified (by ∀ or ∃) is unambiguous and called a *"sentence"*. The first example is not a sentence because x and y are free variables. The last example is a false sentence, but if the universe is the set of real numbers, then it is a true sentence.

In the logic we have described, a variable can represent only an object in the universe. Such a logic is called a *"first-order logic"*, and a theory that uses this logic is called a *"first-order theory"*. For example, in first-order arithmetic, the universe is the set of natural numbers, so any variable can represent *only* a natural number, not a set of numbers nor a predicate. In higher-order logics, variables may represent objects that are not in the universe. First-order logic is powerful enough to handle almost all of Mathematics. *Unless otherwise stated, "theory" in this book will mean "first-order theory".*

So far, we have the language of logic and a proof mechanism. The language includes *logical symbols* such as ∀ and →. The proof mechanism is defined by logical axioms and rules of inference. But mathematical proofs need more than bare logic. For example, to study arithmetic, we first enlarge our vocabulary to include symbols like **0** and **+**. These are not related to logic and are called *"non-logical symbols"*. We then need some axioms about natural numbers, for example, ∀x (x+**0** = x). These are called *"non-logical axioms"*. A *mathematical theory* is defined by its language (i.e. non-logical symbols) and a set of non-logical axioms. These augment the logical symbols and logical axioms, giving us the complete language and proof system of that theory.

Suppose T is a theory with language $\mathcal{L}$. To form a *structure* for $\mathcal{L}$, we first choose a set $\mathbb{U}$ to be the universe of the structure. Objects in the universe are often called "elements". Then we assign a meaning to each non-logical symbol in $\mathcal{L}$. A function on $\mathbb{U}$ is assigned to each function symbol, and an object in $\mathbb{U}$ is assigned to each constant symbol. To each predicate symbol (except =), we assign a predicate on $\mathbb{U}$. If the Equality predicate symbol = is in $\mathcal{L}$, we may not specify what it means; it must mean "__ is the same object as __". Thus, **b** = **c** is true if and only if **b** and **c** are names of the same object. (Note: an object in $\mathbb{U}$ needs to have a name before statements can refer to it. If necessary, more constant symbols may need to be added to $\mathcal{L}$, but we will not go into the technical details.) We now have a structure for $\mathcal{L}$. Any statement in $\mathcal{L}$ can now be interpreted in the structure we have defined. If our choices have been made in such a way that the non-logical axioms of T are all true in our structure, then the structure is called a *"model"* of T.

3.2 Group Theory

We now define *group theory* as an example of a mathematical theory. In Mathematics, *groups* are sets with certain structural properties which are described below. They crop up in many situations, such as the relationships found in solutions of polynomial equations, in particle Physics, in coordinate systems of Special Relativity, and in the configurations of a Rubik's Cube. In the early 19th century, Évariste Galois was one of the first to recognize the significance of groups, and he proved some deep theorems about them. If mathematicians ever decide to name the GOAT (Greatest Of All Time), he would be a strong candidate for the title. He died tragically at the age of 21.

The language of group theory consists of the logical symbols, the Equality predicate =, and two non-logical symbols: a function symbol (∘) and a constant symbol (**e**). The non-logical axioms of group theory are:

G1. ∀x ∀y ∀z ((x ∘ y) ∘ z = x ∘ (y ∘ z))
G2. ∀x (**e** ∘ x = x & x ∘ **e** = x)
G3. ∀x ∃y (x ∘ y = **e** & y ∘ x = **e**)

This completes the definition of group theory. We name the theory "Grp".

Functions are sometimes called "operators". We can think of 5 + 2 = 7 as 5 operating on 2 (from the left) via the addition operator, turning it into 7. This is just a different mental picture, and a different way of describing what a function does.

Axiom G1 says that the function ∘ is associative. In the integers, addition is associative; for example, (6 + 3) + 2 is equal to 6 + (3 + 2). But subtraction is not associative because (6 − 3) − 2 is not equal to 6 − (3 − 2).

Axiom G2 says that the object **e** leaves any object unchanged after operating on it from the left or the right. **e** is called an "identity" element.

Axiom G3 says that every element has an inverse. (By definition of "inverse", operating an element on its inverse from either side yields **e**).

A model of group theory is called a "group". To define a group, we first specify the universe, then choose a function to play the role of ∘, and choose one object in the universe to play the role of **e**. Then we need to verify that G1, G2, and G3 are true in the structure we defined.

Example 1. We choose $\mathbb{R}^+$, the set of positive real numbers, as our universe. We choose multiplication as the function ∘, and 1.0 for the role of **e**. You can check that the axioms are all true in this structure. We call this group "$(\mathbb{R}^+, \times)$".

Example 2. We choose $\mathbb{N}_{12}$, the natural numbers from 1 to 12 (i.e. the numbers on a clock face) as our universe. Then we choose addition mod 12 ("adding hours on a clock") to be ∘. For example, 9 + 5 is 2 (because 5 hours after 9 o'clock is 2 o'clock). We choose 12 as **e**. Again, you can check that the axioms are all true in this structure. We call this group "$(\mathbb{N}_{12}, +)$".

Example 3. We choose all the ways that you can shuffle a deck of 3 cards as our universe. There are six ways, and we next describe and name them. Suppose the deck is now arranged as [T/M/B], which means the Top card sits over the Middle card which sits over the Bottom card. The name of each shuffle is the arrangement that results from doing that shuffle.

There are three shuffles that switch the positions of two cards.
    1. [T/B/M] switches the positions of the Middle and Bottom cards.
    2. [B/M/T] switches the positions of the Top and Bottom cards.
    3. [M/T/B] switches the positions of the Top and Middle cards.

There are two shuffles that move each card to a new position.
    4. [M/B/T] moves each card, with the Middle card ending on top.
    5. [B/T/M] moves each card, with the Bottom card ending on top.

Finally, there is one shuffle that leaves all the cards in their original positions.
    6. [T/M/B]

You can check that these are the only ways that 3 cards can be shuffled. These six shuffles are the objects in our universe. We define the function ∘ as follows: if **b** and **c** are objects in our universe, then **b** ∘ **c** is the shuffle that results when you first apply **c**, and then apply **b** to a deck of three cards. The combined result is a shuffling of the deck, so it must be one of the objects in our universe. This means that we have indeed defined a function because the only requirement for a function is that the output is an object in our universe. The object that we nominate to be **e** is [T/M/B]. Verifying that the axioms are all true in this structure is not difficult. This group is named "S_3". Note that

    [T/B/M] ∘ [M/B/T] is [M/T/B]
but   [M/B/T] ∘ [T/B/M] is [B/M/T],
so   [T/B/M] ∘ [M/B/T] and [M/B/T] ∘ [T/B/M] are different objects. **[3.8]**

Consider this sentence, which we call "*Comm*":

$$\forall x \, \forall y \, (x \circ y = y \circ x).$$

It says that the order of the ∘ operation does not affect the result of the operation. Any such operation is said to be "commutative". In the integers, addition is commutative, but subtraction is not (e.g. $5 - 3 \neq 3 - 5$). A group in which the operator ∘ is commutative is called a "commutative group". In the three groups we considered, the first two are commutative, but [3.8] shows that S_3 is not commutative. *Comm* is true in the first two groups, but false in S_3.

3.3 Introduction to Metamathematics

For formalists, sentences in a formal theory are meaningless strings of symbols. They acquire meaning only when interpreted within a structure, or in other words, only after we specify what the universe is, and what each non-logical symbol stands for. Until then, we literally don't know what we are talking about. For example, the symbols **e** and ∘ have a different meaning in each of our three groups. We may be tempted to say that **e** ∘ **e** = **e** is true, but it does not even have a fixed meaning. In $(\mathbb{R}^+, \times)$, it means 1.0×1.0 is 1.0, but in $(\mathbb{N}_{12}, +)$, it means $12 + 12$ is 12. In S_3, it means that doing the [T/M/B] shuffle twice has the same effect as doing it once. If a sentence *on its own* does not even have a fixed meaning, how can we call it "true"? Our earlier assertion that "every line in a proof is true" should be clarified. We should say that every line in a formal proof is *true in every model*, because the axioms are *true in every model*, and the rules of inference preserve *truth in every model*.

As we defined earlier, a sentence φ is a *theorem* in a theory T (abbreviated by "$T \vdash \varphi$") if φ can be proved from the axioms of T. Note that the symbol ⊢ is not a symbol in the formal language of any theory; it is used as an abbreviation in our informal language, which is English (until this book is translated).

We assign one element to be **e** when defining a group, but are there other elements that behave like an identity? Consider the following sentence (which we name "*eUniq*"):

$$\forall x \, (\exists y \, (x \circ y = y) \to x = \mathbf{e}).$$

eUniq says that if any element in a group behaves like an identity (leaving at least one element unchanged after operating on it from the left), then that element must be **e**. In other words, the identity element is unique. You can check that in each of our three groups, *eUniq* is true. Is it true in every group?

A sentence φ is said to be *"valid"* in a theory T if φ is true in every model of T. The expression "$T \vDash \varphi$" means that φ is valid in T. The symbol ⊨ is an abbreviation in our informal language (English), and is not a symbol in any formal language. We suspect that every group has a unique identity element. In other words, we suspect that Grp$\vDash eUniq$. But there are infinitely many groups, so it is impossible to check each one. How can we be sure that *eUniq* is valid?

Suppose that T is a theory, $\mathfrak{M}$ is any model of T, and φ is a theorem of T. By definition of "model", the axioms of T are all true in $\mathfrak{M}$. The first line in the formal proof of φ is true in $\mathfrak{M}$ because the first line is an axiom. Thus, every line of the proof is true in $\mathfrak{M}$ because the rules of inference preserve truth in $\mathfrak{M}$. Therefore, φ (the last line of the proof) is true in $\mathfrak{M}$. But $\mathfrak{M}$ was any model of T, so φ is true in every model of T. In other words, φ is valid in T, or $T \vDash \varphi$. Therefore, for any theory T and any sentence φ in the language of T,

if $T \vdash \varphi$ then $T \vDash \varphi$.

To summarize, suppose T is a theory and φ is a sentence in the language of T.

1. $T \vdash \varphi$ means that φ is provable in T. It means that the manipulation of meaningless strings of symbols using axioms and rules of inference can produce φ, another meaningless string of symbols. It says nothing about truth.

2. "φ is true in $\mathfrak{M}$" and "φ is false in $\mathfrak{M}$" refer to the Truth or Falsity of φ when it is interpreted in $\mathfrak{M}$, a model of T.

3. $T \vDash \varphi$ means that φ is valid in T. It means φ is true in *every* model of T, and makes no reference to what is provable in T.

4. If $T \vdash \varphi$ then $T \vDash \varphi$. This metamathematical theorem, which says that all theorems are valid, is called the *"Soundness Theorem for First-order Logic"*. It is true because the rules of inference preserve truth in any model. The proof of this theorem is finitary because there are only finitely many rules of inference, so to check that they all preserve truth in any model is a finitary procedure.

If you already knew that (roughly speaking) "theorems are true", then what the Soundness Theorem says is not surprising. It says that first-order logic is sound, meaning it is sensible or reliable. Logic would not be sound if we could prove a sentence that is not valid (i.e. is false in some model).

Now we can determine if every group has a unique identity element. It turns out that *eUniq* is provable in Grp. (Try proving it informally!) Then according to the Soundness Theorem, *eUniq* is valid in **Grp**, which means it is true in every group. Therefore, the identity element is unique in every group.

A theory T is *inconsistent* if there is a sentence φ such that $T \vdash$ φ and $T \vdash \neg$φ. In other words, some sentence and its negation are both provable in T. A *consistent* theory is one that is not inconsistent.

The Soundness Theorem tells us something about theories that have at least one model. We assume that T is a theory with a model $\mathfrak{M}$. We now make a second assumption that T is inconsistent. Then by definition, there is a sentence φ such that $T \vdash$ φ and $T \vdash \neg$φ. According to the Soundness Theorem, this means that $T \vDash$ φ and $T \vDash \neg$φ. In other words, φ and ¬φ are both valid. We conclude that φ and ¬φ are both true in $\mathfrak{M}$. But if one is true, the other is false, so a sentence and its negation cannot both be true in $\mathfrak{M}$. Our second assumption that T is inconsistent has forced us to a false conclusion. Therefore, T cannot be inconsistent. What we have shown is this:

If T is a theory that has a model, then T must be consistent. **[3.9]**

[3.9] gives us a method of proving that a theory T is consistent in two steps.

1. Prove that T has a model.
2. Use [3.9] (the fact that any theory with a model must be consistent).

[3.9] is a simple consequence of the Soundness Theorem, which is provable by finitary methods. Therefore, Step 2 does not violate the condition that our proof should use only finitary methods. But what about Step 1? If T is so complicated that proving the existence of a model involves non-finitary methods, then our whole proof becomes non-finitary. On the other hand, if T is simple enough that we can prove it has a model by finitary methods, then we would have given a finitary proof that T is consistent.

Group theory is simple enough that we can prove it has models by finitary methods. Consider S_3, the group consisting of all 3-card shuffles. Its universe consists of only 6 elements, so Axiom G1 (declaring that ∘ is associative) can be exhaustively confirmed by checking 216 statements (because each of x, y, and z in G1 may be one of only 6 elements). Axioms G2 and G3 can also be exhaustively confirmed by checking only finitely many statements. Therefore, S_3 can be proved to be a model of group theory using only finitary methods. This is a finitary proof that group theory is consistent.

Suppose T is inconsistent. Then for some sentence φ, $T \vdash \varphi$ and $T \vdash \neg\varphi$. If we write these two proofs as one proof, then by the &-Introduction Rule,

$$T \vdash (\varphi \ \& \ \neg\varphi). \quad [3.10]$$

But for *any* sentences φ and θ,

$$(\varphi \ \& \ \neg\varphi) \to \theta$$

is a logical axiom; it is always true in any model because $\varphi \ \& \ \neg\varphi$ is always false. (Recall that $\alpha \to \beta$ is true whenever α is false.) Therefore,

$$T \vdash (\varphi \ \& \ \neg\varphi) \to \theta. \quad [3.11]$$

If we join the proofs in [3.10] and [3.11] into a single proof, and apply the $\to$-Elimination Rule, we conclude that

$$T \vdash \theta.$$

Therefore, *any* sentence is provable in an inconsistent theory. By Soundness, *any* sentence is valid in an inconsistent theory. Any sentence θ is true in every model because an inconsistent theory has no models in which θ is false.

The Soundness Theorem for first-order logic says that

In any first-order theory, every provable sentence is valid,

or, For any first-order theory T and sentence φ, if $T \vdash \varphi$ then $T \vDash \varphi$. [3.12]

If we turn [3.12] around, we get a statement with a different meaning:

In any first-order theory, every valid sentence is provable,

or, For any first-order theory T and sentence φ, if $T \vDash \varphi$ then $T \vdash \varphi$. [3.13]

[3.13] is called the "Completeness Theorem for First-order Logic". Its proof is more complex than the proof of the Soundness Theorem, and we will investigate it in the next chapter.

First-order logic is not the only proof system that can be devised. Other systems may be of higher order, or have different logical axioms, or even different notions of what "true" means. But almost all systems have the notions of provability and validity. We think of a proof system as being "strong" if it can prove lots of theorems. For any logic, we would like it to be strong enough to prove all valid sentences because valid sentences "should" be provable. But we do not want it to be so strong that it can prove invalid sentences because what is provable should never be false in any model. The Completeness Theorem for First-order Logic tells us that first-order logic has all the strength we want; *every* valid sentence is provable. The Soundness Theorem for First-order Logic tells us that first-order logic is not excessively strong; invalid sentences are *never* provable. The strength of first-order logic, like the temperature of Baby Bear's porridge, is just right.

A theory T is said to be *"negation-incomplete"* if there is a sentence φ such that φ and $\neg\varphi$ are both unprovable in T (abbreviated as "$T \nvdash \varphi$" and "$T \nvdash \neg\varphi$"). Such a sentence φ is said to be *"formally undecidable in T"*. In any model, either φ is true or $\neg\varphi$ is true. Therefore, in any model $\mathfrak{M}$ of a negation-incomplete theory T, there is a true sentence that is not provable in T (since φ and $\neg\varphi$ are both unprovable, but one of them is true in $\mathfrak{M}$). A theory is *negation-complete* if it is not negation-incomplete.

For example, recall that *Comm* is false in the group S_3, and $\neg Comm$ is false in the group $(\mathbb{N}_{12}, +)$. By definition of "valid", neither *Comm* nor $\neg Comm$ is valid in group theory. In other words, Grp$\nvDash$*Comm* and Grp$\nvDash\neg$*Comm*. The Soundness Theorem then tells us that neither *Comm* nor $\neg Comm$ is provable in Grp, or in other words, Grp$\nvdash$*Comm* and Grp$\nvdash\neg$*Comm*. This means that *Comm* is a formally undecidable sentence in Grp, which means that Grp is negation-incomplete.

A negation-incomplete theory is one in which the axioms are not extremely restrictive. Any formally undecidable sentence is true in some models and false in others, so the models of a negation-incomplete theory have enough freedom that they can disagree over the truth or falsity of some sentences. Suppose we add another axiom to the three axioms G1, G2, and G3 of Grp:

   G4. *Comm*

Now we have a new theory, and all its models are commutative groups. We call it "ComGrp". With its extra axiom, ComGrp is more powerful and more restrictive than Grp. There are sentences that are formally undecidable in Grp, but not in ComGrp. For example, *Comm* is formally undecidable in Grp but not in ComGrp because it is a theorem in ComGrp (provable in one line!).

Although ComGrp is more restrictive than Grp, ComGrp is still negation-incomplete. For example, consider this sentence:

   $\exists x \exists y \forall z (z = x \lor z = y)$.

It says that there are not more than two objects in the universe. This sentence is true in some commutative groups, and false in others. According to Soundness, it is formally undecidable in ComGrp.

We showed above that there are formally undecidable sentences in Grp, making it negation-incomplete. But in some consistent theories, there are no formally undecidable sentences. For example, consider the theory $\langle S_3 \rangle$ where

> the language of $\langle S_3 \rangle$ is the language of Grp, and
> the axioms of $\langle S_3 \rangle$ are the sentences that are true in S_3.

Then $\langle S_3 \rangle$ is consistent (because S_3 is a model). It is negation-complete because for any sentence φ, either φ is true in S_3 or $\neg\varphi$ is true in S_3, so one of these two sentences is provable in $\langle S_3 \rangle$ because one of them is an axiom of $\langle S_3 \rangle$.

In any theory, there are three types of sentences.

Type 1: Provable sentences

These are the theorems of the theory. They are true in every model (i.e. they are valid). Here are some provable sentences in Grp:

> All the axioms, G1, G2, and G3
> *eUniq*
> $\forall y \, (y = y)$, which is a logical axiom about Equality

Type 2: Refutable sentences

A sentence is *refutable* if its negation is provable. A refutable sentence is false in every model. Here are some refutable sentences in Grp:

> $\neg$G1 (because $\neg\neg$G1 is provable), $\neg$G2, and $\neg$G3
> $\neg eUniq$
> $\exists y \, (\neg y = y)$

Type 3: Formally Undecidable sentences

Formally undecidable sentences are neither provable nor refutable. They are true in some models, and false in others. Here are some formally undecidable sentences in Grp:

> *Comm*
> $\neg Comm$
> $\exists x \, \forall z \, (z = x)$, which says the universe has exactly one object
> $\forall x \, \exists z \, (\neg z = x)$, which says the universe does not have exactly one object

In an inconsistent theory, every sentence is provable, every sentence is refutable, and no sentence is formally undecidable.

CHAPTER 4
THE COMPLETENESS OF FIRST-ORDER LOGIC

4.1 The Question of Completeness

In 1929, Gödel presented the proof of the Completeness Theorem for first-order logic as his doctoral dissertation. Even if he never proved another theorem, his name would be remembered for this highly significant achievement. But this was not a case of a young mathematician brilliantly settling a long-standing unsolved problem; the truth was a little more complicated. We now take a look at the question of Completeness.

The Soundness Theorem for First-order Logic says that

In any first-order theory, every provable sentence is valid,

or, For any first-order theory T and sentence φ, if $T \vdash \varphi$ then $T \vDash \varphi$. [4.1]

The converse of an implication is the statement that interchanges the "if" and "then" parts of the implication. The converse of [4.1] is [4.2] below:

In any first-order theory, every valid sentence is provable,

or, For any first-order theory T and sentence φ, if $T \vDash \varphi$ then $T \vdash \varphi$. [4.2]

[4.2] is called the *"Completeness Theorem for First-order Logic"*. A proof system "should" be able to prove all valid sentences, so one that has this power (e.g. first-order logic) is said to be "complete". We saw [4.1] and [4.2] earlier as [3.12] and [3.13]. The Soundness Theorem must have been known for decades, although not by that name. For example, mathematicians knew that in group theory, every theorem is true in every group, or in other words, that every theorem is valid. You might think it would have been natural to wonder if the converse of Soundness is true, especially when the converse is such a significant statement: logic is powerful enough to prove all valid sentences. Yet the question of Completeness was not among Hilbert's famous twenty-three problems of 1900. Nor was it mentioned when Hilbert introduced the formalist program in 1923. But logicians had not been sitting around waiting for Hilbert to pose problems. In 1920, Emil Post had proved the Completeness Theorem for propositional logic (where the language has no variables or quantifiers). Where was the Completeness Theorem for first-order logic?

In 1928, Hilbert and Wilhelm Ackermann published *Principles of Mathematical Logic*, which laid out in detail the system that we have called "first-order logic". This classic text strengthened the position of first-order logic as the appropriate foundation for Mathematics. Until then, the situation was less clear, although several logicians had proved significant theorems about first-order logic. Students of logic today may have difficulty imagining the situation in the 1920s, a period when important concepts were just coming into clear focus. In *Principles of Mathematical Logic*, Hilbert and Ackermann explicitly posed the question of Completeness for first-order logic.

But according to Dawson, there is no evidence that Gödel found his dissertation topic from Hilbert and Ackermann. Moreover, their book appeared only shortly before Gödel's thesis. Neither, according to Gödel himself, did the topic come from his supervisor Hans Hahn; the dissertation was completed before Hahn saw it. Dawson writes (p. 54), "...*it seems best to suppose that Gödel came upon his dissertation topic on his own...*". Evidently, Gödel was a self-supervising graduate student, both finding and solving his research topic by himself. Faced with his first challenge, he had prevailed alone and without a leader (just as Rumpole would some twenty years later).

Gödel's proof in his dissertation was neither a jewel of innovative ingenuity nor a dazzling display of technical prowess. In 1923, Thoralf Skolem had proved a significant theorem in *model theory*, the study of the relationships between formal languages and their interpretations in structures. Skolem's proof was intricate, involving the manipulation of quantifiers and systematic analysis of the forms of statements. He was not trying to prove the Completeness Theorem for first-order logic, but he came close to doing so. Gödel recognized the significance of Completeness, and saw how close Skolem came to proving it. Then "all" he did was provide the missing steps. Almost four decades later, Gödel wrote that although it was surprising that Skolem and others had missed the proof of Completeness, the prevailing attitudes and state of knowledge at that time had made it difficult to see.

The fact that parts of his proof of the Completeness Theorem were due to Skolem does not diminish Gödel's achievement. On the contrary, it is astonishing that one so young could have peered through murky waters and discerned what was important while more experienced eyes could not. This was the first of many occasions when Gödel's ability to identify problems of significance was revealed. Innovative ingenuity and dazzling technical prowess would be revealed very shortly.

4.2 The Compactness Theorem

Gödel published the proof of the Completeness Theorem in 1930, but what he published was not the proof in his dissertation. Instead, he proved the Compactness Theorem, a surprising theorem that no-one had anticipated, and from it, he derived the Completeness Theorem. These two theorems lie at the heart of model theory. In the opinion of some logicians, Compactness is even more significant than Completeness.

The *Compactness Theorem for First-order Logic* says that

> if every finite subset of a set of first-order sentences has a model, then the whole set has a model.

When the whole set is finite, then the theorem is trivially true (because the whole set is then a finite subset of itself), so Compactness is significant only for infinite sets of sentences.

Gödel posed and answered the question of Compactness for first-order logic on his own. His 1930 publication shed light on the connections and distinctions between the two nascent fields of proof theory and model theory. By the age of 24, Gödel had twice demonstrated his ability to hone in on concepts of crucial importance that were generally not clearly understood, to clarify them, and to prove the pivotal properties that underlie them.

4.3 Completeness, Compactness, and the Model Existence Theorem

The following metamathematical theorem, earlier seen as [3.9], is a simple consequence of the Soundness Theorem:

> If T is a theory that has a model, then T must be consistent. **[4.3]**

Just as the converse of the Soundness Theorem is the Completeness Theorem, the converse of [4.3] is also an important theorem known as the "*Model Existence Theorem for First-order Logic*". As the converse of [4.3], it says

> If T is a consistent first-order theory, then T has a model. **[4.4]**

Gödel was aware of the connection between the Completeness Theorem, the Compactness Theorem, and the Model Existence Theorem. Any one of them can be proved without much difficulty by assuming any one of the other two. What is difficult is to prove any of them "from scratch", without assuming either of the other two. We will take only a superficial look at the proof of the Completeness Theorem.

4.4 Gödel's Proof of the Completeness Theorem

Gödel later wrote that *"The Completeness Theorem is indeed an almost trivial consequence of ...* [Skolem's proof]", and added that the inference involved non-finitary reasoning. Thus, his proof of the Completeness Theorem is not finitary, and he knew this was unavoidable. Although the statement of the theorem is metamathematical in nature, its proof fails to meet the condition set down by Hilbert. Looking for a finitary proof is pointless because the Completeness Theorem implies other theorems that are known to be non-finitary in nature.

To show you an example of non-finitary reasoning, we now state *König's Lemma*, which belongs to a branch of Mathematics known as "graph theory". When Gödel modified Skolem's proof to prove the Completeness Theorem, he assumed König's Lemma at one step of the proof, and this assumption made his proof non-finitary. The diagram below shows a *binary tree*. Each vertex is shown as a circle. There is one vertex at the top that is connected to any other vertex by exactly one path. (By definition, a "path" must start at the top and move down.) At each vertex, a path may terminate, or continue as one path, or split into two paths. (If you think this "tree" looks more like a root system, turn it upside-down, and it will at least look like a bush.)

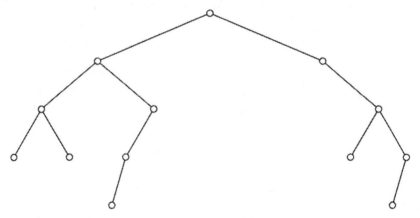

In this tree, 5 paths join more than 3 vertices, and 2 paths join more than 4 vertices. König's Lemma states that if a binary tree has arbitrarily long paths, then it has at least one path that joins infinitely many vertices. (Having "arbitrarily long paths" means that for any number you name, the tree has a path that joins more vertices than the number you named.)

To some people, König's Lemma is "obviously true". Yet it cannot be proved from axioms that are universally accepted in graph theory, set theory, or any other theory. One axiom that does enable you to prove König's Lemma is the Axiom of Choice, the controversial axiom mentioned in Chapter 2 that is definitely non-finitary. Although König's Lemma is "less non-finitary" than the Axiom of Choice, it is still questionable enough to make Gödel's proof of the Completeness Theorem non-finitary. Any proof of the Completeness, Compactness, or Model Existence Theorems requires either König's Lemma or some other non-finitary assumption.

Dénes König proved the lemma that bears his name in 1927 using a weak form of the Axiom of Choice. Generally speaking, a lemma is a theorem that is not considered significant enough to bear the title of "theorem". Lemmas are often useful as stepping stones to prove other theorems.

Those unfamiliar with metamathematical proofs may find it difficult to imagine how something like König's Lemma could be used in a proof of Completeness. All we can say by way of explanation is that structures found in metamathematics occur also in other branches of Mathematics. Equally puzzling might be how the Compactness Theorem was named. It is closely related to Tychonoff's Theorem, an important theorem in topology concerning *compact* topological spaces.

4.5 The "Completeness Theorem" with a Finitary Proof

When proving the statement (If **A** then **B**), we may assume **A** "for free" (i.e. without any justification), and then use the assumption to prove **B**. But the "free" assumption has a price; when we have proved **B**, all we can claim to have proved is (If **A** then **B**), and not the stronger statement **B**.

So, to prove the following version of the Completeness Theorem,

If König's Lemma is true, then first-order logic is Complete [4.5]

we may assume König's Lemma. As a result, our proof of [4.5] (i.e. the deduction taking us from König's Lemma to the conclusion that first-order logic is Complete) is finitary. By hiding the assumption, the Completeness Theorem [4.2] superficially appears to be a stronger theorem than [4.5]. But [4.5] is no weaker, and by explicitly revealing its non-finitary assumption, it is more informative.

CHAPTER 5
ARITHMETIC

5.1 Axiomatizing Arithmetic

Arithmetic is the branch of Mathematics that studies the natural numbers, 0, 1, 2,...etc., together with the functions of addition and multiplication. The set of natural numbers (denoted by "$\mathbb{N}$") is infinite, and interest in the foundations of arithmetic grew amidst the controversy surrounding Cantor's work on infinite sets, and the earlier one over non-Euclidean geometry. A rigorous foundation for arithmetic was sought, and the way forward was to find a set of axioms that would turn arithmetic into an axiomatic theory. There was also much interest in the foundations of analysis, the study of real numbers. Despite being far more complex than natural numbers, Dedekind had shown that real numbers could be defined in terms of natural numbers. Thus, the foundations of analysis rested on arithmetic, making arithmetic the natural place to start when investigating the integrity of the foundations of Mathematics.

The goal of axiomatizing arithmetic was different from the goal of axiomatizing other theories such as group theory. The axioms of **Grp** are very unrestrictive, allowing for the existence of many different types of groups. Each theorem is true in all groups, so we may say that "each theorem proves many truths". With arithmetic, there was one "intended" model for the axioms: the natural numbers. The hope was that axioms could be found that have the intended model as their *only* model, that can be proved to be consistent, and that are strong enough to make all the true sentences of arithmetic provable. *The best laid schemes o' Mice and Men ...*

A set of axioms presented by Giuseppe Peano in 1889 succeeded in capturing precisely the natural numbers; the only model (called "$\mathfrak{N}$") was the intended one. But one variable in one axiom (called the "Induction Axiom") stood for a subset of the universe, rather than an element. This axiom is thus a sentence of second-order logic, making Peano's theory a second-order theory. Some logicians are critical of second-order logic because its strength is not "just right". The eminent logician W.V. Quine criticized it as "set theory in disguise". To many logicians, the only proper logic for Mathematics is first-order logic. But the Induction Axiom appeared essential; without it, many sentences that are true in $\mathfrak{N}$ cannot be proven. We now take a look at the Peano Axioms.

5.2 The Peano Axioms

The Peano axioms are meant to capture arithmetic on the natural numbers, which includes the functions of addition and multiplication. Therefore, the language of this set of axioms must include the function symbols + and ×. When we build the intended model, we take ℕ as our universe, and the functions of addition and multiplication will be assigned to + and ×. We also add the constant symbol **0** to the language, and assign to it the number zero. To enable our axioms to capture the properties of + and ×, and to name the rest of the natural numbers in the language, we add another function symbol **s** to the language. This function symbol takes only one input, and when we build the intended model, we assign the successor function to **s**. (Given any natural number as input, the *successor function* outputs the next higher natural number. For example, successor(25) is 26.) Let "$\mathcal{L}_{PA}$" denote the language with the Equality predicate symbol (=) and four non-logical symbols: **0**, **s**, +, and ×. Then to start off, Peano essentially chose these six axioms:

P1. $\forall u\, (\, \neg\, \mathbf{s}(u) = \mathbf{0}\,)$
P2. $\forall u\, \forall v\, (\, \mathbf{s}(u) = \mathbf{s}(v) \rightarrow u = v\,)$
P3. $\forall u\, (\, u + \mathbf{0} = u\,)$
P4. $\forall u\, \forall v\, (\, u + \mathbf{s}(v) = \mathbf{s}(u + v)\,)$
P5. $\forall u\, (\, u \times \mathbf{0} = \mathbf{0}\,)$
P6. $\forall u\, \forall v\, (\, u \times \mathbf{s}(v) = (u \times v) + u\,)$

You can check that in our intended model, all six axioms are true, so $\mathfrak{N}$ is a model of P1 to P6. It is called the *"standard model"*, and its universe is ℕ. A model of P1 to P6 where the universe is not ℕ is called a *"non-standard model"*.

We now show that the universe of *any* model of P1 to P6 contains the natural numbers. Suppose $\mathfrak{M}$ is a model of P1 to P6, and 𝕄 is its universe. Some object in 𝕄 is assigned to be **0**; we refer to it as "Z". Some function on 𝕄 is assigned to be **s**; we refer to it as "F". Consider the infinite list [5.1] below:

$\qquad$ Z, F(Z), F(F(Z)), F(F(F(Z))),...etc. $\qquad\qquad$ [5.1]

[5.1] is a list of objects in 𝕄 because F is a function. In the model $\mathfrak{M}$, the predicate symbol = of $\mathcal{L}_{PA}$ must mean "__ is the same object as __". Therefore, P2 says that if J and K are any objects in 𝕄 such that

$\qquad$ F(J) and F(K) are the same object $\qquad\qquad$ [5.2]
then
$\qquad$ J and K are the same object. $\qquad\qquad$ [5.3]

In other words, whenever [5.2] is true, we may "cancel" F from both sides of "and" in [5.2] to give us [5.3], and [5.3] is also true.

Now suppose that two objects listed in [5.1] are the same object. Then

$F(F(F(F(...(Z))))...)$ and $F(F(F(...(Z)))...)$ are the same object [5.4]

where the number of Fs on the two sides of "and" are different. [5.4] has the same form as [5.2], so we may cancel F from both sides of "and" in [5.4]. By cancelling as many Fs as possible, we reach the conclusion that

F(something) and Z are the same object. [5.5]

But F was assigned to be **s**, and Z assigned to be **0**. So [5.5] shows that P1 is false in $\mathfrak{M}$. This cannot happen because $\mathfrak{M}$ is a model of P1 to P6. Therefore, our assumption that two of the objects listed in [5.1] are the same must be incorrect. We conclude that [5.1] is a list of objects in $\mathbb{M}$ that never repeats. Since Axioms P3 to P6 are true in $\mathfrak{M}$, these objects behave under + and × just like the natural numbers behave. Therefore, we can think of the objects in list [5.1] as 0, 1, 2, 3, ... etc. We have shown that the universe of any model of P1 to P6 contains a copy of the natural numbers, or we can just say that

the universe of any model of P1 to P6 contains the natural numbers, and the names of 0, 1, 2, 3,... etc. are **0, s(0), s(s(0)), s(s(s(0)))**,... etc. [5.6]

We abbreviate **0, s(0), s(s(0)), s(s(s(0)))**,...etc. by $\bar{0}, \bar{1}, \bar{2}, \bar{3}$,...etc. Then for any number n, $\bar{n}$ is an expression in $\mathcal{L}_{PA}$ that names n. $\bar{n}$ is called a *"numeral"*. For example, $\bar{2}$ abbreviates **s(s(0))**; it is the numeral that names the number 2. Note that $\bar{2}$ (or **s(s(0))**) is a word in the language $\mathcal{L}_{PA}$, but 2 is not.

Axioms P1 to P6 allow too many models. For example, we can build a model of P1 to P6 called "$\mathfrak{Q}^+$" by choosing $\mathbb{Q}^+$ (the set of non-negative fractions) as the universe, and making the obvious choices for the four symbols **0, s, +**, and ×. You can verify that P1 to P6 are all true in $\mathfrak{Q}^+$. As expected, $\mathfrak{Q}^+$ contains the natural numbers (each natural number is a non-negative fraction). Why don't we want $\mathfrak{Q}^+$ as a model? Consider this sentence:

$\neg \exists x \; x + x = \mathbf{s(0)}$ [5.7]

[5.7] says that there is nothing in the universe that when added to itself gives the answer 1. [5.7] is true in $\mathfrak{N}$, and we certainly want it to be provable. But [5.7] is false in $\mathfrak{Q}^+$ because 0.5 is in its universe. This means that in the theory of P1 to P6, [5.7] is not a valid sentence, and therefore [5.7] cannot be proved using these six axioms. $\mathfrak{Q}^+$ is a non-standard model of P1 to P6.

ARITHMETIC

Peano needed to add an axiom that allows only natural numbers to be in the universe. So he added the *Induction Axiom*:

> For every subset $\mathbb{S}$ of the universe, if [$\mathbb{S}$ includes 0, and includes the successor of anything in $\mathbb{S}$], then $\mathbb{S}$ is the whole universe. [5.8]

P1 to P6 ensure that the universe of any model includes the natural numbers, so $\mathbb{N}$ is a subset of the universe of any model. Moreover, $\mathbb{N}$ includes 0 and the successor of anything in $\mathbb{N}$. [5.8] then says that $\mathbb{N}$ is the whole universe. Therefore, the universe of any model of P1 to P6 and [5.8] is exactly $\mathbb{N}$, so $\mathfrak{N}$ is the only model of these axioms. [5.8] is a second-order sentence because $\mathbb{S}$ stands for a subset of the universe. So the theory with P1 to P6 and [5.8] as axioms is known as "*Second-order Peano Arithmetic*", and denoted by "PA2".

5.3 First-order Peano Arithmetic

PA2 succeeded in the sense that the standard model is its only model. But as explained earlier, we prefer to work in first-order logic. So we replace the Induction Axiom by a first-order axiom P7 that we hope will do what the Induction Axiom did (i.e. exclude all non-standard models). P7 is a set of infinitely many first-order sentences, called the "first-order induction axiom schema". Each individual axiom in P7 is called a "P7-axiom". Here is P7:

> P7. If $\varphi(x)$ is any (first-order) statement in $\mathcal{L}_{PA}$, and x is its only free variable, then [5.9] below is an axiom:
>
> $$[\ \varphi(0)\ \&\ \forall k\ (\ \varphi(k) \to \varphi(s(k))\ )\ ] \to \forall x\ \varphi(x). \qquad [5.9]$$

Peano Arithmetic (denoted by "PA") is the theory with $\mathcal{L}_{PA}$ as its language, and P1 to P7 as its axioms. PA is a first-order theory.

Note: $\varphi(0)$, $\varphi(k)$, and $\varphi(s(k))$ are the statements we get as the result of replacing every x in $\varphi(x)$ by **0**, k, and **s**(k) respectively.

For example, when $\varphi(x)$ is [5.10] below:

$$\neg\ x + x = s(0) \qquad [5.10]$$

then [5.9] becomes [5.11] below:

$$[\ \neg 0+0=s(0)\ \&\ \forall k\ (\ \neg k+k=s(0) \to \neg s(k)+s(k)=s(0)\ )\ ] \to \forall x\ \neg x+x=s(0) \qquad [5.11]$$

[5.11] is an axiom of PA. It is one of infinitely many P7-axioms.

Note that [5.9] is not an axiom of PA. It is not even a sentence in $\mathcal{L}_{PA}$ because φ is not in the vocabulary of $\mathcal{L}_{PA}$. [5.9] is a pattern that generates all the P7-axioms as φ ranges over all statements in $\mathcal{L}_{PA}$ with one free variable.

In Appendix A, we show that $\mathfrak{N}$ is a model of PA (the theory with P1 to P7 as axioms), and $\mathfrak{Q}^+$ is not. Thus, Axiom P7 succeeds in excluding $\mathfrak{Q}^+$ as a model, but as we will see below, it does not exclude all non-standard models.

Although there are infinitely many P7-axioms, we can tell if any given sentence is a P7-axiom or not by using a routine procedure that is guaranteed to end. We just need to check the structure of the given sentence, and see if it matches the structure of [5.9]. We do not have to check the given sentence against an infinite list, looking for a match; such a procedure might never end. This is an important point because if we cannot effectively tell what the axioms of a theory are, then we cannot tell if a given sequence of statements is a proof or not. That would not be a useful theory.

5.4 A Non-Standard Model of Peano Arithmetic

We now show that PA has a non-standard model. First, we define a new theory which we call "PA$^\infty$". We start with PA, and extend its language $\mathcal{L}_{PA}$ by adding a new constant symbol ∞. Next, we add axiom P8, an axiom schema that consists of infinitely many first-order axioms.

P8. For every natural number n, the following is an axiom: $\neg \infty = \bar{n}$.

Thus, the P8-axioms are:

$$\neg \infty = \bar{0}, \ \neg \infty = \bar{1}, \ \neg \infty = \bar{2}, \ \neg \infty = \bar{3}, \ \neg \infty = \bar{4}, \ldots \text{etc.} \qquad [5.12]$$

or $\quad \neg \infty = 0, \ \neg \infty = s(0), \ \neg \infty = s(s(0)), \ \neg \infty = s(s(s(0))), \ldots$etc.

Axiom P8 says that the object assigned to ∞ is not a natural number.

Now suppose F is any finite subset of the axioms of PA$^\infty$. Then only finitely many P8-axioms are in F, which means that infinitely many of the sentences in [5.12] are not in F. We need only one. Suppose $\neg \infty = \overline{82}$ is not in F. Then we can make $\mathfrak{N}$ a model of F by assigning 82 to the role of ∞. (Every sentence in F is true in $\mathfrak{N}$ because assigning 82 to ∞ does not violate any of the P8-axioms that are in F, and the other sentences in F are all axioms of PA.)

PA$^\infty$ is a first-order theory, and we have just shown that every finite subset of its axioms has a model. Then according to the Compactness Theorem, the set of all its axioms has a model. In other words, PA$^\infty$ has a model. We call it

"$\mathfrak{N}^\infty$". Being a model of PA$^\infty$, there must be an object in the universe of $\mathfrak{N}^\infty$ that is assigned to the role of ∞. But Axiom P8 forbids this object to be a natural number. Therefore, some object in the universe of $\mathfrak{N}^\infty$ is not a natural number. Since $\mathfrak{N}^\infty$ is a model of PA$^\infty$, it is also a model of PA (because every axiom of PA is an axiom of PA$^\infty$). With an object in its universe that is not a natural number, $\mathfrak{N}^\infty$ is a non-standard model of PA.

The existence of non-standard models of PA is not caused by P7 being a "bad choice" as an axiom. Suppose T is *any* first-order theory that has $\mathcal{L}_{PA}$ as its language, and has $\mathfrak{N}$ as a model. Then T *must* have non-standard models. To show this, we do the same thing to T that we just did to PA. We add a new constant symbol ∞ to the language and add P8 as a new axiom. Then the Compactness Theorem leads us to a non-standard model of T. Therefore, no first-order axiom or axiom schema could have excluded all non-standard models. This deduction demonstrates the power of the Compactness Theorem.

5.5 The Compactness Theorem Fails for Second-order Logic

We now take a short detour, and show that the Compactness Theorem does not hold in second-order logic. We start with PA2, second-order Peano Arithmetic. Then, as we did before, we add a new constant symbol ∞ to its language, and add all the P8-axioms below

$$\neg \infty = \overline{0}, \ \neg \infty = \overline{1}, \ \neg \infty = \overline{2}, \ \neg \infty = \overline{3}, \ \neg \infty = \overline{4}, \ \ldots \text{etc.} \quad [5.12]$$

as new axioms. This gives us a new second-order theory that we call "PA2$^\infty$". As before, $\mathfrak{N}$ is a model of any finite subset of the axioms of PA2$^\infty$ because some P8-axioms must be missing from any finite subset. Therefore, every finite subset of the axioms of PA2$^\infty$ has a model. But the entire set of axioms of PA2$^\infty$ has no models because the Induction Axiom says that every object in the universe is a natural number, while P8 says that at least one object is not a natural number. Thus, the Compactness Theorem fails for at least one second-order theory.

5.6 Peano Arithmetic and Set Theory

In Section 3.4, it was claimed that ZFC (set theory with the Axiom of Choice) is powerful enough to incorporate essentially all of Mathematics, which includes PA (Peano Arithmetic). But its language $\mathcal{L}_{ZF}$ does not have any of the non-logical symbols of $\mathcal{L}_{PA}$, so statements in $\mathcal{L}_{PA}$ are not even statements in $\mathcal{L}_{ZF}$.

So in what sense does ZFC incorporate (or contain) PA? It turns out that the universe of any model of ZFC contains elements that behave exactly like natural numbers. "Behave" means that functions can be defined in ZFC that have the same effect on "ZFC-numerals" that addition and multiplication have on numerals in PA. Thus, hiding within every model of ZFC is a copy of the natural numbers, together with functions exactly like addition and multiplication. Every statement in $\mathcal{L}_{PA}$ can be translated to a statement of $\mathcal{L}_{ZF}$ with the same meaning, and the axioms of PA (when translated) are all provable in ZFC. Thus, every theorem of PA (when translated) is a theorem of ZFC. In that sense, PA is part of set theory. But set theory is more powerful than PA; there are sentences in $\mathcal{L}_{PA}$ that are unprovable in PA, but are provable in ZFC (after being translated). An example of such a sentence is given in Section 8.4. See Appendix B for more information on how the natural numbers may be represented in set theory.

There is no point in asking if the numbers in models of ZFC are the "same things" as those in models of PA. The essential point is that the objects we think of as "Zero", "One", "Two", etc. are found in every model of PA and of ZFC. Moreover, in any model of either theory, these objects are related to each other in exactly the same way that the natural numbers are related under addition and multiplication. From a mathematical point of view, any set of objects that behave exactly like the natural numbers *are* the natural numbers.

5.7 Arithmetic in 1930

In 1930, two central problems remained unsolved: to prove that (axiomatic) arithmetic is consistent, and to prove that all true sentences of arithmetic are provable. Partial successes had already been achieved. In 1925, Ackermann had proved the consistency of a theory without quantifiers in its language. His work was later extended by von Neumann. In 1929, two simplified versions of first-order Peano Arithmetic had been shown to be consistent and negation-complete, implying that all true sentences in models of these theories are provable. One theory was due to Mojżesz Presburger, and the other to Skolem. By 1930, Gödel's Completeness and Compactness Theorems had clarified important concepts, and were two new weapons in the logicians' arsenal.

Nobody thought the two central problems would be easily solved, but the stage appeared set for the decisive theorems to be proved. In 1931, decisive theorems were indeed proven. However, they were not the expected ones.

CHAPTER 6
FORMAL UNDECIDABILITY IN ARITHMETIC

6.1 The "Incompleteness" Theorems

At the Conference on Epistemology of the Exact Sciences held in Könisberg in September 1930, Gödel quietly mentioned during a general discussion session that there are true sentences of arithmetic that are not formally provable. This unexpected announcement should have caused a sensation, but according to Dawson (p. 69), there is no record of any reaction. The announcement may have passed under the radar because it was not on the official agenda, or more likely because it was neither fully understood nor believed. Among those present, von Neumann may have been the only one to grasp the significance of what Gödel had said, and they later conversed privately.

In 1931, the proof of Gödel's claim was published under the title *"On Formally Undecidable Propositions of Principia Mathematica and Related Systems I"*. In this paper (probably intended as the first of two), he also sketched a proof of a second theorem with even greater significance: the consistency of a formal system of arithmetic cannot be proved within that system. The two theorems are known as the "First and Second Incompleteness Theorems of Arithmetic". The two central problems had been solved, but solved "negatively" in one brilliant publication that many mathematicians struggled to understand.

Before we examine the theorems, two points should be clarified. The first concerns their common name: the "Incompleteness Theorems". We have already seen the Completeness Theorem for first-order logic. Unfortunately, the "completeness" of first-order logic is unrelated to the "incompleteness" of arithmetic. This ambiguity in the word "complete" could lead to confusion; the theorems do not say that arithmetic lacks something that first-order logic possesses. What Gödel proved in 1931 (which is clearly indicated by the title of his paper) is that there are formally undecidable sentences in formal systems of arithmetic. A theory with such sentences is called "negation-incomplete", so Gödel's theorems of 1931 should be called the "Negation-Incompleteness Theorems of Arithmetic". But the incorrect name is now firmly established (at least in English), and it may sound better than the correct name. So we will also say (reluctantly) that a theory with formally undecidable sentences is "incomplete" rather than "negation-incomplete". *If you can't beat 'em, ...*

The second point to clarify concerns *"Principia Mathematica and Related Systems"*. Gödel did more than prove that one particular formal system of arithmetic is incomplete, or that it cannot prove its own consistency. That would have been shocking enough, but his proof shows that these limitations apply to *any* reasonable formal system of arithmetic. ("Reasonable" will be clarified later.) The sweeping generality of the Incompleteness Theorems greatly increases their significance. For example, true unprovable sentences of arithmetic occur not only in *Principia Mathematica* (*PM*), but in several *"Related Systems"* as well. One of them is Zermelo-Fraenkel set theory (ZFC). The fact that such sentences of arithmetic occur in both *PM* and ZFC is especially significant because they are very powerful theories; if they cannot prove all the truths of arithmetic, then this limitation is very deeply rooted. Another related system is first-order Peano Arithmetic (PA). It too is incomplete, and it is easier to work with than *PM* or ZFC. So we will examine the Incompleteness Theorems as though they refer to PA; what they reveal about PA applies also to *"Principia Mathematica and Related Systems"*.

6.2 The First Incompleteness Theorem

The First Incompleteness Theorem applies to several different formal systems. When applied to Peano Arithmetic (PA with language $\mathcal{L}_{PA}$), it says that

    if PA is consistent, then PA is *incomplete*, meaning that
        there is a sentence in $\mathcal{L}_{PA}$ that is formally undecidable in PA,
or   for some sentence G in $\mathcal{L}_{PA}$, PA$\nvdash G$ and PA$\nvdash \neg G$.   [6.4]

Note that the theorem does not simply say that some sentence in $\mathcal{L}_{PA}$ is formally undecidable in PA. It says that such a sentence exists *if* PA is consistent. (There is a minor wrinkle here about the assumption of consistency that will be ironed out later.) For the theorem to be a proper metamathematical one, it should have a finitary proof. But its proof relied on the consistency of PA which (as we will see later) cannot be proved by finitary methods. Therefore, the consistency of PA had to be explicitly stated as an assumption.

Gödel's proof of the First Incompleteness Theorem is finitary and constructive. The existence of the sentence G in [6.4] was proved by directly constructing it, step by step. Then Gödel gave a finitary proof that both G and $\neg G$ are unprovable in PA if PA is consistent. But one of them must be true in $\mathfrak{N}$ (and it happens to be G). A sentence in arithmetic is said to be "true" if it is true in $\mathfrak{N}$, the intended model of arithmetic. Therefore, if PA is consistent, then there is a sentence in arithmetic that is true but unprovable in PA. This was essentially what Gödel announced at the conference in 1930.

In theories with no intended model, incompleteness is not very significant. For example, Grp is incomplete because *Comm* is a formally undecidable sentence in Grp. *Comm* is true in some groups and false in others, and the same goes for ¬*Comm*. But there is no "intended model" of group theory, so neither sentence can be considered to be simply "true". Thus, although group theory is incomplete, there are no "true unprovable sentences" in group theory.

At first glance, there seems to be a way of eliminating the discomfort of having a sentence G that is true in $\mathfrak{N}$ but unprovable in PA. We simply add G as a new axiom, joining the other axioms of PA to create a new theory called "PA+G". In the theory PA+G, G is provable (because it is an axiom). The discomfort of having a true unprovable sentence has been easily removed. But the same discomfort promptly re-emerges. The proof of the Incompleteness Theorem describes a procedure that uses the power of PA to produce a sentence G that is true in $\mathfrak{N}$, but unprovable in PA (assuming that PA is consistent). But PA+G is even more powerful than PA, and if PA is consistent, so is PA+G (because $\mathfrak{N}$ is a model). So the same procedure can be applied to the theory PA+G to produce a sentence, a "new G", that is true in $\mathfrak{N}$, but unprovable in PA+G (assuming that PA is consistent). We are still in the same situation. True unprovable sentences appear to be unavoidable in reasonable theories of arithmetic. Thus, the First Incompleteness Theorem was an unexpected setback for the formalist program although at first, this was not widely recognized.

6.3 Proof of the First Incompleteness Theorem

The proof of the First Incompleteness Theorem is long and difficult, and we will barely scratch its surface. But at its core, almost hidden beneath layers of technical complexities, the proof has an elegant simplicity.

The proof may be divided into two separate tasks:

Task 1.   Construct the sentence G.
Task 2.   Show that PA$\nvdash G$ and PA$\nvdash \neg G$ (assuming PA is consistent).

First, we will look at an overview of the proof. In the overview, we will give only a description of G (i.e. what it says), and then describe an "easy" way of doing Task 2 (which is not entirely satisfactory). Those who understand the overview can claim to have some insight into Gödel's proof. But the devil is in the details, so we will then take a closer look at the structure of G, and the role of the ingenious technique of Gödel numbering. Then we can examine how Gödel completed Task 1 in greater detail. Finally, we will describe the better way in which Gödel completed Task 2.

6.4 An Overview of Gödel's Proof

We now describe the sentence G without attempting to show how it was constructed. G is related to the Liar Paradox sentence, which says that

This sentence is not true. [6.5]

[6.5] says of itself that it is not true. If [6.5] is true, then what it says (about itself) is correct. Then according to itself, it is *not* true. If [6.5] is not true, then what it says (about itself) is incorrect. Then according to itself, it *is* true. But [6.5] is either true or not true, and either way, there is a contradiction.

Gödel modified the Liar Paradox, constructing a sentence G in $\mathcal{L}_{PA}$ that says

This sentence is not provable. [6.6]

By saying "not provable" instead of "not true", [6.6] avoids the contradiction of [6.5]. Instead, [6.6] turns out to be formally undecidable. Here is a quick and very informal way to do Task 2. If [6.6] is false, then what it says is incorrect, which means it is provable. Thus, [6.6] is a false provable sentence, which violates Soundness. Therefore, [6.6] cannot be false. This means that what it says is correct, so [6.6] is not provable. Thus, [6.6] is a true unprovable sentence. The negation of [6.6] is false, so by Soundness, it too is unprovable. Therefore, [6.6] (or G) is formally undecidable and true.

To do the overview more carefully, we first express [6.6] more precisely:

G is true in $\mathfrak{N}$ if and only if G is not provable in PA, [6.7]

or $\quad \neg G$ is true in $\mathfrak{N}$ if and only if G is provable in PA. [6.8]

Next, we do Task 2: show that $PA \nvdash G$ and $PA \nvdash \neg G$. We assume $\mathfrak{N}$ is a model of PA (which means we are also assuming that PA is consistent). G is either true in $\mathfrak{N}$ or false in $\mathfrak{N}$, so either Case 1 or Case 2 below must be correct.

Case 1. G is true in $\mathfrak{N}$ (and $\neg G$ is false in $\mathfrak{N}$)
Since G is true in $\mathfrak{N}$, then by its very meaning (see [6.7]), G is not provable in PA. The sentence $\neg G$ is false in $\mathfrak{N}$, so by the Soundness Theorem, it cannot be provable in PA. Therefore, we have $PA \nvdash G$, and $PA \nvdash \neg G$, and G is true in $\mathfrak{N}$.

Case 2. G is false in $\mathfrak{N}$ (and $\neg G$ is true in $\mathfrak{N}$)
Since $\neg G$ is true in $\mathfrak{N}$, then by its very meaning (see [6.8]), G is provable in PA. But G is false in $\mathfrak{N}$, a model of PA. By the Soundness Theorem, a provable sentence cannot be false in any model, so Case 2 is impossible.

With Case 2 shown to be impossible, Case 1 must be correct. Therefore, if $\mathfrak{N}$ is a model of PA, then G is formally undecidable in PA, and G is true in $\mathfrak{N}$.

In this "easy" way of doing Task 2, we used the Soundness Theorem and the "very meaning" of G. Gödel also did Task 2 this way in the opening section of his paper to give the reader an intuitive grasp of the proof. The proof that he gave in the main body of his paper did not rely on Soundness or on the meaning of G. That made his proof stronger and more acceptable. Soundness involves the notions of "truth" and "validity" which were viewed with suspicion by many logicians, being more nebulous than the concrete and finitary notion of a proof. Moreover, the statement of the First Incompleteness Theorem mentions only consistency and formal undecidability. Both these concepts are defined in terms of proofs; they are unrelated to the model-theoretic concepts of Soundness and truth. If Gödel had completed *only* the easy version of Task 2, his proof would have been criticised.

6.5 Constructing "G in English"

We now take a small detour and construct a sentence in English that asserts its own unprovability. The process is amusing, and gives us some insight into how Gödel constructed the sentence G in $\mathcal{L}_{PA}$. This example is taken from Hartley Rogers, Jr., "*Theory of Recursive Functions and Effective Computability*", McGraw-Hill, 1967.

The construction of "G in English" involves the substitution of one expression within another. Specifically, if **P** and **Q** are expressions, then by definition,

The result of substituting "**P**" *in* "**Q**"

is the expression **Q** in which every occurrence of X has been replaced by **P**.

For example,

The result of substituting "*23*" *in* "*85 + X^2 = X^3 + X*"   [6.9]

is   *85 + 23^2 = 23^3 + 23*   [6.10]

[6.9] is an expression giving a precise, accurate description of [6.10], and the *only* thing it describes is [6.10]. Therefore, [6.10] *is* [6.9]. They are the "same thing" just as Half of ten is the "same thing" as Five. If we want to say that [6.10] is not provable, we can say it by placing the words *"The following is not provable:"* in front of [6.9]:

The following is not provable: The result of substituting
"23" in "$85 + X^2 = X^3 + X$" [6.11]

[6.11] says that [6.9] is not provable. But since [6.9] and [6.10] are the "same thing", [6.11] also says that [6.10] is not provable. By saying something about [6.9], [6.11] is a statement that also says something about [6.10].

We now repeat what we just did with a more complicated expression:

The result of substituting
"The following is not provable: The result of substituting "X" in "X"" in
"The following is not provable: The result of substituting "X" in "X"" [6.12]

Making the substitution specified by [6.12] (and you should check this for yourself), we have [6.13] below:

The following is not provable: The result of substituting
"The following is not provable: The result of substituting "X" in "X"" in
"The following is not provable: The result of substituting "X" in "X"" [6.13]

As in the previous example, [6.13] *is* [6.12]; they are the "same thing". If we want to say that [6.13] is not provable, we can say it by placing the words *"The following is not provable:"* in front of [6.12], giving us [6.14]:

The following is not provable: The result of substituting
"The following is not provable: The result of substituting "X" in "X"" in
"The following is not provable: The result of substituting "X" in "X"" [6.14]

As in the previous example, [6.14] says that [6.12] is not provable. But since [6.12] and [6.13] are the "same thing", [6.14] also says that [6.13] is not provable. But [6.13] is identically the same expression as [6.14]. So [6.14] says that [6.14] is not provable. [6.14] is a sentence that asserts its own unprovability. What we just did in English, Gödel did in the language of PA.

6.6 Gödel Numbering

Gödel numbering is an ingenious technique that is widely used in proofs today. Using it, Gödel constructed G, a sentence in $\mathcal{L}_{PA}$ with this property:

G is true in $\mathfrak{N}$ if and only if G is not provable in PA. [6.7]

FORMAL UNDECIDABILITY IN ARITHMETIC

Putting aside for now the self-referential nature of G, the first hurdle for us is to see how a sentence in $\mathcal{L}_{PA}$ (a language designed to talk about numbers) can assert that some sentence is unprovable. First, we need some definitions.

For any statement φ in $\mathcal{L}_{PA}$ that has one free variable, we let "φ($\bar{n}$)" denote the sentence we get by replacing every occurrence of the free variable by the numeral $\bar{n}$. For example, if φ is

$$\exists u\,(\,s(u + u) = w \times w\,) \qquad [6.15]$$

then φ is a statement with one free variable w, and φ($\bar{3}$) is this sentence:

$$\exists u\,(\,s(u + u) = s(s(s(0))) \times s(s(s(0)))\,) \qquad [6.16]$$

We sometimes write φ as φ(w) to remind ourselves that w is the only free variable in φ. We name [6.15] as "*Od*" or "*Od*(w)". Then [6.16] is *Od*($\bar{3}$).

Numbers have many properties, such as the property of being odd. A property P is said to be *expressible* in $\mathcal{L}_{PA}$ if there is a statement θ in $\mathcal{L}_{PA}$ such that

for any number n, θ($\bar{n}$) is true in $\mathfrak{N}$ if and only if n has property P.

In such a case, we say that θ($\bar{n}$) *asserts* that n has the property P, and we say that P is *expressed by* θ.

For example, *Od*(w) says that the successor of some even number is equal to the square of w. This is a round-about way of saying that w is odd. Thus, the property of being an odd number is expressed by the statement *Od* because

for any number n, *Od*($\bar{n}$) is true in $\mathfrak{N}$ if and only if n is odd,

or   for any number n, *Od*($\bar{n}$) asserts that n is odd.

For example, *Od*($\bar{3}$) is true in $\mathfrak{N}$, and *Od*($\bar{4}$) is false in $\mathfrak{N}$.

These definitions extend to properties involving more than one number. If we abbreviate $\exists u\,(\,x + s(u) = y\,)$ by "*LesThan*(x, y)", then

for any numbers m and n, *LesThan*($\bar{m}$, $\bar{n}$) asserts that m is less than n.

Statements in $\mathcal{L}_{PA}$ can make assertions only about numbers. But Gödel wanted to use $\mathcal{L}_{PA}$ to make assertions about statements in $\mathcal{L}_{PA}$. So he devised a scheme to assign a unique number to each statement (called its "*Gödel number*"). For any statement φ, we let "⌜φ⌝" denote its Gödel number. Properties of φ are captured as numerical properties of ⌜φ⌝. By referring to properties of ⌜φ⌝, statements in $\mathcal{L}_{PA}$ can indirectly say things about φ. He also assigned Gödel numbers to finite sequences of statements so that $\mathcal{L}_{PA}$ can make assertions about them too. Think of Gödel numbers as "translations" of statements and finite sequences of statements in $\mathcal{L}_{PA}$ into a "language" that consists only of numbers. Then ⌜φ⌝ is just the statement φ expressed in another language. $\mathcal{L}_{PA}$ can make assertions about the statement φ by making assertions about its translation ⌜φ⌝.

If n is a Gödel number, we let "$\boxed{n}$" denote the object with Gödel number n. The object $\boxed{n}$ is either a single statement or a finite sequence of statements.

Gödel's scheme of assigning Gödel numbers has two key properties:

1. Every bit of information about the structure of a statement or a finite sequence of statements is contained in its Gödel number (i.e. its translation).

2. All this information can be recovered from a Gödel number n by simple arithmetical operations that translate n back to $\boxed{n}$.

For example, if φ is a statement that has the form of an implication $\alpha \to \beta$, this fact can be revealed by analysing its Gödel number ⌜φ⌝, and can be expressed in $\mathcal{L}_{PA}$. This means there is a statement *Imply*(w) in $\mathcal{L}_{PA}$ such that for any number n, *Imply*($\bar{n}$) is true in $\mathfrak{N}$ if and only if $\boxed{n}$ has the form of $\alpha \to \beta$.

The Gödel numbering system is artificial. Every symbol in $\mathcal{L}_{PA}$ is coded by a number. Suppose the symbol $\to$ happens to be coded by the number 3. Then within *Imply*($\bar{n}$), the following phrase can be found: $= s(s(s(0)))$. This part of *Imply*($\bar{n}$) says that something equals 3. This must be said somewhere in *Imply*($\bar{n}$) because *Imply*($\bar{n}$) must say that $\to$ is part of $\boxed{n}$, and it refers to $\to$ as $s(s(s(0)))$. ($\mathcal{L}_{PA}$ cannot say that something "$= \to$".) But there is no reason why $\to$ should be coded by 3. To those who don't know the coding system, *Imply*($\bar{n}$) is a sentence that says for no apparent reason that something is equal to 3 (among many other equally pointless assertions).

Gödel showed that many properties of (Gödel numbers of) statements and finite sequences of statements are expressible in $\mathcal{L}_{PA}$. Using them, he constructed a crucially important statement *PrfPA*(x, w) in $\mathcal{L}_{PA}$ with this property:

for any numbers m and n, *PrfPA*($\bar{m}$, $\bar{n}$) is true in $\mathfrak{N}$
if and only if $\boxed{m}$ is a proof in PA of $\boxed{n}$. [6.17]

To construct *PrfPA*, Gödel showed that the property of being a proof is expressible in $\mathcal{L}_{PA}$. This meant showing that the properties of "being an axiom of PA" and "following the rules of inference" are expressible in $\mathcal{L}_{PA}$. This was not easy. For us, a rough guideline is that any property that is verifiable in a finite number of routine steps is expressible in $\mathcal{L}_{PA}$. This is why we made the point in Section 5.3 that whether or not a sentence is a P7-axiom can be verified

FORMAL UNDECIDABILITY IN ARITHMETIC   45

in a finite number of routine steps. If this were not so, then "being an axiom" would not be an expressible property, and the statement *PrfPA* would not exist.

Summary of our notation: for all n and φ,

    n (in sans serif font) is a variable in $\mathcal{L}_{PA}$, the formal language of PA,
    n (in *italics*) is a variable in English that stands for a number,
    $\bar{n}$ is a numeral, a string of symbols in $\mathcal{L}_{PA}$ that names the number *n*,
    ⌜φ⌝ is the Gödel number of the statement φ,
    ⟦n⟧ is the object with Gödel number *n* (so ⌜⟦n⟧⌝ is *n*, and ⟦⌜φ⌝⟧ is φ).

Gödel craftily constructed *G* so that its Gödel number ⌜*G*⌝ occurs within itself.

    G is the sentence ¬∃x *PrfPA*(x, $\overline{⌜G⌝}$).   [6.18]

(We will indicate in Section 6.7 how he did this.) According to [6.17],

    G is true in 𝔑 if and only if
    there is no number *m* such that ⟦m⟧ is a proof in PA of the sentence ⟦⌜*G*⌝⟧.

But ⟦⌜*G*⌝⟧ is *G*. Therefore,

    G is true in 𝔑 if and only if *G* is not provable in PA.   [6.19]

By including its own Gödel number within itself, *G* is a self-referential sentence. *G* is a sentence in $\mathcal{L}_{PA}$ that asserts its own unprovability.

If you are satisfied with our overview of Gödel's proof (Section 6.4) and the above description of *G*, you can skip Sections 6.7 and 6.8 without great loss.

6.7 Task 1: The Construction of *G*

We now describe how Gödel constructed *G* as specified by [6.18], and how he managed to include the Gödel number of *G* within itself. Take the particular variable z. We call *b* a "*z-Gödel number*" if *b* is the Gödel number of a statement with z as its *only* free variable. When *b* is a z-Gödel number, ⟦b⟧ will be denoted by "ψ$_b$(z)". Therefore, $b = $ ⌜ψ$_b$(z)⌝.

For example, suppose the Gödel number of z = s(z) is 523. Then

    523 is ⌜z = s(z)⌝
    523 is a z-Gödel number (because z is the only free variable in z = s(z))
    ⟦523⟧ is this statement: z = s(z) (by definition of ⟦☐⟧)
    ψ$_{523}$(z) is this statement: z = s(z) (by definition of ψ$_{523}$(z))
    523 is ⌜ψ$_{523}$(z)⌝ (because of the previous line)
    ψ$_{523}$($\bar{2}$) is this sentence: $\bar{2} = s(\bar{2})$, which is s(s(0)) = s(s(s(0))).

A *term* is an expression in $\mathcal{L}_{PA}$ that names a number when all the free variables within it are replaced by numerals. For example, suppose $Dble(y)$ is the expression y+y. Then $Dble(y)$ is a term, and $Dble(\overline{4})$ is $\overline{4}+\overline{4}$, which names 8.

Gödel built a term that we have simplified and called "$zSub(y)$". The term $zSub(y)$ has the following property:

if b is a z-Gödel number, $zSub(\overline{b})$ names the Gödel number of
($\boxed{b}$ after all free occurrences of z in $\boxed{b}$ have been replaced by $\overline{b}$).

When b is a z-Gödel number, $\boxed{b}$ is $\psi_b(z)$, and $zSub(\overline{b})$ names ⌜$\psi_b(\overline{b})$⌝.

Building $zSub(y)$ is not easy. We can believe it exists because what it "does" involves three finite routine processes: recovering $\boxed{b}$ from b, replacing every z in $\boxed{b}$ by $\overline{b}$ to form a new statement, and expressing the Gödel number of the new statement. Such processes are expressible $\mathcal{L}_{PA}$.

For example, suppose again that the Gödel number of $z = s(z)$ is 523. Then

$\psi_{523}(z)$ is this statement: $z = s(z)$

$\psi_{523}(\overline{523})$ is this sentence: $\overline{523} = s(\overline{523})$

$zSub(\overline{523})$ names ⌜$\psi_{523}(\overline{523})$⌝, which is the Gödel number of $\psi_{523}(\overline{523})$.

Consider this statement:

$PrfPA(x, zSub(y))$. [6.20]

[6.20] is $PrfPA(x, w)$ after every free occurrence of w has been replaced by $zSub(y)$. As a result, [6.20] has two free variables, x and y. According to [6.17],

for any numbers m and b, if b is a z-Gödel number,
$PrfPA(\overline{m}, zSub(\overline{b}))$ asserts that $\boxed{m}$ is a proof in PA of $\psi_b(\overline{b})$ [6.21]

(because $zSub(\overline{b})$ names ⌜$\psi_b(\overline{b})$⌝). From [6.21], we conclude that

for any number b, if b is a z-Gödel number,
¬∃x $PrfPA(x, zSub(\overline{b}))$ asserts that $\psi_b(\overline{b})$ is not provable in PA [6.22]

(because no object $\boxed{m}$ is a proof of $\psi_b(\overline{b})$). Now consider the statement

¬∃x $PrfPA(x, zSub(z))$ [6.23]

Let "**g**" denote the Gödel number of [6.23]. (Note: a **bold** letter stands for one particular fixed number; **g** is not a variable in $\mathcal{L}_{PA}$ or in English.) In the special case when b is **g**, [6.22] tells us that

if **g** is a z-Gödel number,
¬∃x $PrfPA(x, zSub(\overline{\mathbf{g}}))$ asserts that $\psi_\mathbf{g}(\overline{\mathbf{g}})$ is not provable in PA. [6.24]

But [6.23] is a statement with z as its only free variable. Therefore, its Gödel number **g** is a z-Gödel number. Then according to [6.24],

$\neg \exists x \, PrfPA(x, zSub(\overline{g}))$ asserts that $\psi_g(\overline{g})$ is not provable in PA. [6.25]

Since **g** is a z-Gödel number, then by definition, $\psi_g(z)$ is the statement with Gödel number **g**. But the statement with Gödel number **g** is [6.23]. Therefore, $\psi_g(z)$ is [6.23], which is $\neg \exists x \, PrfPA(x, zSub(z))$, and therefore,

$\psi_g(\overline{g})$ is $\neg \exists x \, PrfPA(x, zSub(\overline{g}))$. [6.26]

Note that $\psi_g(z)$ is not a sentence because it has a free variable z. But $\overline{g}$ is a numeral, so $\psi_g(\overline{g})$ is a sentence. Combining [6.25] and [6.26],

$\psi_g(\overline{g})$ asserts that $\psi_g(\overline{g})$ is not provable in PA.

We name $\psi_g(\overline{g})$ as "G". Then G is a sentence in $\mathcal{L}_{PA}$ such that

G asserts that G is not provable in PA,

or G is true in $\mathfrak{N}$ if and only if G is not provable in PA,

or G expresses the informal statement that PA$\nvdash G$. [6.27]

Note that $zSub(\overline{g})$ names $\ulcorner \psi_g(\overline{g}) \urcorner$, which is $\ulcorner G \urcorner$. So according to [6.26],

G is the sentence $\neg \exists x \, PrfPA(x, \overline{\ulcorner G \urcorner})$.

Thus, G is the sentence specified in [6.18] that includes its own Gödel number $\ulcorner G \urcorner$ within itself. This completes our description of Task 1.

6.8 Task 2: Proving Formal Undecidability of G in PA

We now describe how Gödel completed Task 2 without using the meaning of G (i.e. its interpretation in $\mathfrak{N}$) or appealing to Soundness. First, Gödel proved that if any sentence φ is provable in PA, then PA can also *prove that* φ *is provable in* PA. This means that when $\boxed{m}$ is a proof of $\boxed{n}$, then $PrfPA(\overline{m}, \overline{n})$ is not merely true in $\mathfrak{N}$, but *provable* in PA. In other words, he proved that

for any m and n,

when $\boxed{m}$ is a proof in PA of $\boxed{n}$, then PA$\vdash PrfPA(\overline{m}, \overline{n})$. [6.28]

Note that there are three proofs involved here: proof in PA of $\boxed{n}$, proof in PA of $PrfPA(\overline{m}, \overline{n})$, and the informal proof of [6.28]. Proving [6.28] is difficult. For our purposes, we just need a reason to believe it. [6.28] is believable because when $\boxed{m}$ is a proof in PA of $\boxed{n}$, this fact can be verified in a finite number of routine steps, which implies that any sentence in $\mathcal{L}_{PA}$ expressing this fact can be proved in PA. One such sentence is $PrfPA(\overline{m}, \overline{n})$.

We now show that if PA is consistent, then PA⊬G. To do this, we assume that PA⊢G, and prove that PA must then be inconsistent. Since PA⊢G, then for some number **m**, $\boxed{m}$ is a proof of G. So by [6.28],

PA⊢*PrfPA*($\overline{m}$, $\ulcorner G \urcorner$). [6.29]

([6.29] says that PA can prove that $\boxed{m}$ is a proof in PA of G.)
Applying the ∃-Introduction Rule to [6.29], we can conclude that

PA⊢∃x *PrfPA*(x, $\ulcorner G \urcorner$).

But we assumed that PA⊢G, and G is ¬∃x *PrfPA*(x, $\ulcorner G \urcorner$). In other words,

PA⊢¬∃x *PrfPA*(x, $\ulcorner G \urcorner$).

Therefore the sentence ∃x *PrfPA*(x, $\ulcorner G \urcorner$) and its negation are both provable in PA, so PA is inconsistent. Therefore, if PA is consistent, then our assumption that PA⊢G must be wrong. Therefore, if PA is consistent, then PA⊬G.

To prove that PA⊬¬G, Gödel had to assume a slightly stronger version of consistency called "ω-consistency" (pronounced as "omega-consistency"). This is the wrinkle mentioned in Section 6.2. PA is said to be "ω-*inconsistent*" if there is a statement φ(x) in $\mathcal{L}_{PA}$ such that the sentences in the infinite list [6.30] are *all provable* in PA. (Note that the first one is different from the rest.)

∃x φ(x),   ¬φ($\overline{0}$), ¬φ($\overline{1}$), ¬φ($\overline{2}$), ¬φ($\overline{3}$), ..., ¬φ($\overline{n}$), ...etc. [6.30]

We say that PA is ω-*consistent* if it is not ω-inconsistent.

Now we show that if PA is ω-consistent, then PA⊬¬G. We begin by assuming that PA is ω-consistent. Then for any statement φ(x), at least one sentence in list [6.30] is unprovable. Therefore, PA is consistent (because in an inconsistent theory, *every* sentence is provable). We already showed that if PA is consistent, then PA⊬G. Therefore, PA⊬G, so for any number *m*, $\boxed{m}$ is *not* a proof of G. This means that for any *m*, ¬*PrfPA*($\overline{m}$, $\ulcorner G \urcorner$) is *provable* in PA (because the fact that $\boxed{m}$ is not a proof of G can be verified in a finite number of routine steps). In other words, the sentences in [6.31] below are *all provable* in PA.

¬*PrfPA*($\overline{0}$,$\ulcorner G \urcorner$), ¬*PrfPA*($\overline{1}$,$\ulcorner G \urcorner$), ¬*PrfPA*($\overline{2}$,$\ulcorner G \urcorner$),.., ¬*PrfPA*($\overline{n}$,$\ulcorner G \urcorner$), ..etc. [6.31]

[6.31] is almost [6.30] when φ(x) is *PrfPA*(x, $\ulcorner G \urcorner$); only the first sentence ∃x φ(x), which is ∃x *PrfPA*(x, $\ulcorner G \urcorner$), is missing. Suppose the missing sentence ∃x *PrfPA*(x, $\ulcorner G \urcorner$) is provable in PA. Then the entire list [6.30] is provable when φ(x) is *PrfPA*(x, $\ulcorner G \urcorner$), contradicting our assumption that PA is ω-consistent. Therefore, ∃x *PrfPA*(x, $\ulcorner G \urcorner$) is not provable in PA.

In other words,

$\text{PA} \vdash \exists x \, \mathit{PrfPA}(x, \ulcorner G \urcorner)$, which implies that

$\text{PA} \nvdash \neg \neg \exists x \, \mathit{PrfPA}(x, \ulcorner G \urcorner)$ (because two negations cancel each other).

But $\neg \exists x \, \mathit{PrfPA}(x, \ulcorner G \urcorner)$ is G. Therefore, $\text{PA} \nvdash \neg G$. We have shown that if PA is ω-consistent, then $\text{PA} \nvdash \neg G$. This completes our description of how Gödel completed Task 2 without using Soundness or the meaning of G.

The sentences in [6.30] cannot all be true in $\mathfrak{N}$ because if $\neg \varphi(\bar{n})$ is true in $\mathfrak{N}$ for every n, then $\exists x \, \varphi(x)$ is false in $\mathfrak{N}$. But if PA is ω-inconsistent, every sentence in [6.30] is provable, so some sentence that is false in $\mathfrak{N}$ is provable in PA. Thus, according to Soundness, *if* PA *is ω-inconsistent*, $\mathfrak{N}$ *would not be a model of* PA. This would be a huge surprise, and precipitate a crisis in Mathematics. Therefore, having to make the stronger assumption that PA is ω-consistent is only a minor weakness of Gödel's proof. This became a moot point in 1936 when J. Barkley Rosser modified G to obtain a formally undecidable sentence assuming only that PA is consistent (without assuming ω-consistency).

6.9 The Scope of the First Incompleteness Theorem

To state a more general version of the First Incompleteness Theorem, we roughly identify the properties of PA that were used in its proof.

Consider a theory T that satisfies C1, C2, and C3 in [6.32] below:

C1. the language of T (denoted by "$\mathcal{L}_T$") together with its axioms are able to "capture the meaning" of the axioms of PA,

C2. when translated into $\mathcal{L}_T$, every axiom of PA is a theorem of T,

C3. the property of being the Gödel number of an axiom of T is expressible in $\mathcal{L}_T$. (Such a theory is said to be "*effective*".) **[6.32]**

In Section 5.6, we stated without explanation that ZFC satisfies C1 and C2. We now clarify C1 and C2 by a simple example. Suppose $\mathcal{L}_T$ is similar to $\mathcal{L}_\text{PA}$, but does not have the function symbol **s**. Instead, $\mathcal{L}_T$ has another constant symbol **1**, and the axioms of T force the object named by **1** in any model to behave like the number One. Any statement of $\mathcal{L}_\text{PA}$ with the symbol **s** can be translated to a statement of $\mathcal{L}_T$ that uses **+1**. Then C2 says, for example, that since

$\forall u \, (\neg \mathbf{s}(u) = \mathbf{0})$ is an axiom of PA, then
$\forall u \, (\neg (u+\mathbf{1}) = \mathbf{0})$ is a theorem of T.

The proof of the First Incompleteness Theorem depends on PA being powerful enough to prove certain sentences, for example [6.28]. If the Theorem is to hold for T, then T needs to be at least as expressive and powerful as PA. If we simply said "Every theorem of PA is a theorem of T", then T would be powerful enough, but we would be excluding many suitably powerful theories (such as ZFC) simply because a difference in vocabulary prevents a theorem of PA from being a theorem of T. This is why C1 and C2 need to mention "capture the meaning" and "translation". They say (admittedly in an imprecise way) that T is at least as expressive and powerful as PA without requiring that $\mathcal{L}_T$ and $\mathcal{L}_{PA}$ be the same language.

Condition C3 implies that there is a statement $PrfT(x, w)$ in $\mathcal{L}_T$ such that
for any numbers m and n,
when $\boxed{m}$ is a proof in T of $\boxed{n}$, then $T \vdash PrfT(\overline{m}, \overline{n})$.

$PrfT$ does for T what $PrfPA$ did for PA. Using $PrfT$, and following what we described for PA, a sentence in $\mathcal{L}_T$ can be constructed that asserts its own unprovability. This sentence is formally undecidable in T if T is consistent.

We can now state the First Incompleteness Theorem in a more general setting. Suppose T is a theory satisfying Conditions C1 to C3 in [6.32]. Then (using Rosser's sentence) if T is consistent, there is a sentence G_T in $\mathcal{L}_T$ such that
$T \nvdash G_T$ and $T \nvdash \neg G_T$ (i.e. T is incomplete), implying that in any model of T, there are sentences that are true in that model, but unprovable in T.

Any theory of arithmetic should satisfy C1 and C2 in order to refer to numbers, addition, and multiplication. Roughly speaking, a theory satisfies C3 if we "know what its axioms are". So demanding that a theory of arithmetic be consistent and satisfy C1, C2, and C3 is a reasonable demand; it is asking for very little. We call such a theory a *"reasonable theory of arithmetic"*. Therefore, in any reasonable theory of arithmetic, there are unprovable sentences that are true (i.e. true in $\mathfrak{N}$). This conclusion is not about any particular axiom system, and therein lies the power of Gödel's theorem. The First Incompleteness Theorem applies to second-order theories of arithmetic, and to very powerful systems like *Principia Mathematica* and set theory because they all satisfy C1, C2, and C3. We will next meet two consistent negation-complete theories that are less powerful than PA with less expressive languages. Despite being less powerful, each of these theories can prove any sentence that is true in any of its models.

6.10 Three Negation-Complete Theories of "Arithmetic"

We look at three theories of "arithmetic" that are consistent and negation-complete. The first two were mentioned in Section 5.7. Since the First Incompleteness Theorem does not hold for any of them, none of them are reasonable theories of arithmetic. Therefore, they all fail to satisfy at least one of the three conditions in [6.32].

The first theory was introduced by Presburger in 1929, and is known as "Presburger Arithmetic". The language of this theory is $\mathcal{L}_{PA}$ without the multiplication symbol ×. The axioms are all the axioms of PA that do not have the × symbol. That means dropping P5 and P6, as well as many P7-axioms. We call the theory "A^+", the "Arithmetic of +". Presburger proved that A^+ is consistent and negation-complete. This means that

for every sentence φ in the language of A^+, either $A^+ \vdash \varphi$ or $A^+ \vdash \neg\varphi$.

Now suppose that φ is true in some model of A^+. Then $\neg\varphi$ is not provable in A^+ (because of Soundness), so φ must be provable in A^+. Therefore, any sentence in the language of A^+ that is true in any model of A^+ must be provable in A^+. This is true for any negation-complete theory.

The second theory, known as "Skolem Arithmetic" was introduced by Skolem, also in 1929, and we call it "$A^\times$". Its language is $\mathcal{L}_{PA}$ without the addition symbol +. Its axioms are the axioms of PA that do not include the symbol +. He proved that $A^\times$ is consistent and negation-complete, implying that any sentence in the language of $A^\times$ that is true in any model of $A^\times$ is provable in $A^\times$.

Neither A^+ nor $A^\times$ satisfy any of the three conditions in [6.32]. Their languages are not expressive enough to handle Gödel numbering, and there is no sentence in the language of either theory that asserts its own unprovability. But in 1929, nobody had thought of Gödel numbering, or of modifying the Liar Paradox. (Nobody except for one person, of course.) With two consistent negation-complete theories that were "almost" PA, hopes must have been high in 1929 that the consistency and negation-completeness of PA would soon be proven. With techniques in its proof that were both intricate and revolutionary, the First Incompleteness Theorem was always going to meet some resistance. The resistance may have been increased by the fact that long-held hopes, apparently close to being met, were now dashed. Zermelo was a strong critic of the Theorem, and he was not the only one. Very few mathematicians have the ability and the open-mindedness of von Neumann.

The third negation-complete theory related to PA is the theory where the language is $\mathcal{L}_{PA}$ (the language of PA), and the axioms are all the sentences of $\mathcal{L}_{PA}$ that are true in $\mathfrak{N}$. We call this theory "$\langle\mathfrak{N}\rangle$". For any sentence φ, either φ is true in $\mathfrak{N}$, or $\neg\varphi$ is true in $\mathfrak{N}$. Therefore, one of these two sentences is an axiom of $\langle\mathfrak{N}\rangle$, and hence, one of them is provable in $\langle\mathfrak{N}\rangle$. In other words, $\langle\mathfrak{N}\rangle$ is negation-complete. Any true sentence of arithmetic is provable in $\langle\mathfrak{N}\rangle$ because it is an axiom of $\langle\mathfrak{N}\rangle$, but we do not know which sentences are axioms of $\langle\mathfrak{N}\rangle$. As a mathematical theory for proving theorems, $\langle\mathfrak{N}\rangle$ is useless.

6.11 Expressing Truth in Arithmetic

We now use $\langle\mathfrak{N}\rangle$ to show that truth in $\mathfrak{N}$ is not expressible in $\mathcal{L}_{PA}$. ($\langle\mathfrak{N}\rangle$ is not totally useless!) Suppose $\mathfrak{N}$ is a model of PA. Then $\langle\mathfrak{N}\rangle$ satisfies C1 and C2 in [6.32], and $\langle\mathfrak{N}\rangle$ is consistent (because $\mathfrak{N}$ is a model). But $\langle\mathfrak{N}\rangle$ is negation-complete, so the First Incompleteness Theorem does not hold for $\langle\mathfrak{N}\rangle$. Therefore, $\langle\mathfrak{N}\rangle$ does not satisfy C3 in [6.32]. This means that

if $\mathfrak{N}$ is a model of PA, then the property of being the Gödel number of an axiom of $\langle\mathfrak{N}\rangle$ is not expressible in $\mathcal{L}_{PA}$.

Being an axiom of $\langle\mathfrak{N}\rangle$ means being true in $\mathfrak{N}$ (or simply "true"). Therefore,

if $\mathfrak{N}$ is a model of PA, then the property of being the Gödel number of a true sentence of $\mathcal{L}_{PA}$ is not expressible in $\mathcal{L}_{PA}$. [6.33]

or if $\mathfrak{N}$ is a model of PA, then there is no statement $Tru(w)$ in $\mathcal{L}_{PA}$ such that for any number n, $Tru(\bar{n})$ asserts that $\boxed{n}$ is true in $\mathfrak{N}$.

[6.33] says that if $\mathfrak{N}$ is a model of PA then truth in $\mathfrak{N}$ is not expressible in $\mathcal{L}_{PA}$.

The assumption that $\mathfrak{N}$ is a model of PA is not necessary if we approach the question directly instead of using the First Incompleteness Theorem. [6.34] below does not assume that $\mathfrak{N}$ is a model of PA.

The property of being the Gödel number of a true sentence of $\mathcal{L}_{PA}$ is not expressible in $\mathcal{L}_{PA}$. [6.34]

To prove [6.34], we assume that [6.34] is false, and we will reach a contradiction. (If you skipped Section 6.7, then the remainder of this section will be difficult, but at least you have seen a "proof" of [6.33], the "easy" version of [6.34].) By the definition of "expressible in $\mathcal{L}_{PA}$",

there is a statement $Tru(w)$ in $\mathcal{L}_{PA}$ with the following property:
for any number n, $Tru(\bar{n})$ asserts that $\boxed{n}$ is true in $\mathfrak{N}$. [6.35]

FORMAL UNDECIDABILITY IN ARITHMETIC   53

Consider this statement: $Tru(zSub(y))$. If b is a z-Gödel number, $zSub(\bar{b})$ names ⌜$\psi_b(\bar{b})$⌝. Therefore, [6.35] says that

for any number b, if b is a z-Gödel number,
$Tru(zSub(\bar{b}))$ asserts that $\psi_b(\bar{b})$ is true in $\mathfrak{N}$.   **[6.36]**

From [6.36], we know that

for any number b, if b is a z-Gödel number,
$\neg Tru(zSub(\bar{b}))$ asserts that $\psi_b(\bar{b})$ is not true in $\mathfrak{N}$.   **[6.37]**

Now consider the statement

$\neg Tru(zSub(z))$.   **[6.38]**

Let "**d**" denote the Gödel number of [6.38]. In the special case when b is **d**, [6.37] tells us that

if **d** is a z-Gödel number,
$\neg Tru(zSub(\bar{d}))$ asserts that $\psi_d(\bar{d})$ is not true in $\mathfrak{N}$.   **[6.39]**

But [6.38] is a statement with z as its only free variable. Therefore, its Gödel number **d** is a z-Gödel number. Then according to [6.39],

$\neg Tru(zSub(\bar{d}))$ asserts that $\psi_d(\bar{d})$ is not true in $\mathfrak{N}$.   **[6.40]**

Since **d** is a z-Gödel number, then by definition, $\psi_d(z)$ is $\boxed{d}$. This means that

$\psi_d(z)$ is [6.38], which is $\neg Tru(zSub(z))$, and therefore,

$\psi_d(\bar{d})$ is $\neg Tru(zSub(\bar{d}))$.

Then according to [6.40], $\neg Tru(zSub(\bar{d}))$ asserts that $\neg Tru(zSub(\bar{d}))$ is not true in $\mathfrak{N}$. In other words, $\neg Tru(zSub(\bar{d}))$ is true in $\mathfrak{N}$ if and only if it is not. This is a contradiction. Our assumption that [6.34] is false must be wrong. Therefore, truth in $\mathfrak{N}$ is not expressible in $\mathcal{L}_{PA}$; there is no statement Tru in $\mathcal{L}_{PA}$ that satisfies [6.35], and the sentence $\neg Tru(zSub(\bar{d}))$ does not exist.

6.12 Tarski's Theorem

[6.34] applies not only to $\mathcal{L}_{PA}$, but to the language of any reasonable theory of arithmetic. Compare the two paragraphs below.

1. Gödel showed that in any theory of arithmetic that satisfies the three conditions of [6.32], provability in the theory is expressible. From there, he constructed a sentence that asserted its own unprovability. Assuming consistency, this sentence was shown to be formally undecidable.

2. We assumed that truth is expressible, and from there, we constructed a sentence that asserted its own falsity. This was the Liar Paradox sentence, and led to a contradiction. We conclude that truth is not expressible.

Gödel knew in 1931 that truth in $\mathfrak{N}$ is not expressible. But in the mathematical climate of the time, the notion of objective "mathematical truth" was viewed with deep suspicion by many logicians. Consequently, Gödel, always cautious and reluctant to face controversy, chose not to publish his discovery. In 1933, Alfred Tarski independently made the same discovery, which he published. He expected a hostile reaction, and that's exactly what he got. The theorem below, stating that truth in $\mathfrak{N}$ is inexpressible, is known as "Tarski's Theorem".

Suppose $\mathcal{L}$ is the language of a reasonable theory of arithmetic.

The property of being a true sentence of $\mathcal{L}$ is not expressible in $\mathcal{L}$.

More explicitly,

There is no statement $Tru(x)$ in $\mathcal{L}$ with the following property:
for every number m,
$\boxed{m}$ is a sentence of $\mathcal{L}$ that is true in $\mathfrak{N}$ if and only if $Tru(\overline{m})$ is true in $\mathfrak{N}$.

In other words,

There is no statement $Tru(x)$ in $\mathcal{L}$ with the following property:
for every sentence φ in $\mathcal{L}$,
φ is true in $\mathfrak{N}$ if and only if $Tru(\overline{\ulcorner\varphi\urcorner})$ is true in $\mathfrak{N}$.

If T is any reasonable theory of arithmetic, the property of being the Gödel number of a theorem of T is simple enough to be expressed by a statement in the language of T (namely, by the statement $\exists x\, PrfT(x, y)$). But this is not so for the property of being the Gödel number of a sentence that is true in $\mathfrak{N}$. Being true in $\mathfrak{N}$ is evidently a more elusive property than being provable.

According to Gödel, the telling reason for the existence of true unprovable sentences in any reasonable theory of arithmetic lies in the fact that provability is expressible but truth is not. The explanation is simple:

Since   "true" and "provable" describe different sets of sentences,
and   provable sentences are all true (according to Soundness),
then   true sentences cannot all be provable.

Therefore, true unprovable sentences must exist in any reasonable theory of arithmetic. In a nutshell, this explains the incompleteness of any reasonable theory of arithmetic.

6.13 Diagonalization

"Diagonalization" is the name of a technique first used by Cantor to prove his ground-breaking theorems in set theory. The technique was used by Gödel to set up the self-referential situation we saw in the construction G, a sentence that asserts something about itself. So we have already seen diagonalization in action, but the "diagonal" aspect was not emphasised.

Diagonalization was also used in Section 6.5, where the substitution specified in [6.12] created [6.13], a self-referential sentence in English that asserted its own unprovability:

The result of substituting
"The following is not provable: The result of substituting "X" in "X"" in
"The following is not provable: The result of substituting "X" in "X"" **[6.12]**

[6.12] has this structure:

The result of substituting "something" in "something"

where the two *"something"*'s are identical. By substituting *something* into itself, we obtained a self-referential sentence.

In Section 6.7, we denoted the Gödel number of $\neg \exists x\, PrfPA(x, zSub(z))$ as "**g**", and abbreviated it as "$\psi_g(z)$". Then we replaced every z in $\psi_g(z)$ by $\overline{\mathbf{g}}$, giving us the sentence $\psi_g(\overline{\mathbf{g}})$ that asserted its own unprovability. If we identify a statement with its Gödel number, what we did was substitute "**g** into itself". Again, this resulted in a self-referential sentence. In Section 6.11, we denoted the Gödel number of $\neg Tru(zSub(z))$ as "**d**", and substituted "**d** into itself". This gave us a self-referential sentence asserting its own falsity.

If we place all the sentences $\psi_m(\overline{n})$ in an infinite table with m going down and n going across, the upper-left portion of the table would look like this:

$\psi_0(\overline{0})$	$\psi_0(\overline{1})$	$\psi_0(\overline{2})$	$\psi_0(\overline{3})$	$\psi_0(\overline{4})$	⇒
$\psi_1(\overline{0})$	$\psi_1(\overline{1})$	$\psi_1(\overline{2})$	$\psi_1(\overline{3})$	$\psi_1(\overline{4})$	
$\psi_2(\overline{0})$	$\psi_2(\overline{1})$	$\psi_2(\overline{2})$	$\psi_2(\overline{3})$	$\psi_2(\overline{4})$	
$\psi_3(\overline{0})$	$\psi_3(\overline{1})$	$\psi_3(\overline{2})$	$\psi_3(\overline{3})$	$\psi_3(\overline{4})$	
$\psi_4(\overline{0})$	$\psi_4(\overline{1})$	$\psi_4(\overline{2})$	$\psi_4(\overline{3})$	$\psi_4(\overline{4})$	

⇓ (placed before the last row)

$\psi_g(\overline{\mathbf{g}})$ lies on the diagonal going from the upper-left corner down to the lower-right, from $\psi_0(\overline{0})$ to $\psi_4(\overline{4})$ and beyond. This is why the technique of substituting "something into itself" is called "diagonalization".

6.14 The Completeness Theorem Fails for Second-order Logic

We now show that the Completeness Theorem does not hold for second-order logic. In other words, we show there is a second-order theory in which not every valid sentence is provable.

We assume that second-order Peano Arithmetic (called "PA2") is consistent and that $\mathfrak{N}$ is a model. One of the axioms of PA2 is the Induction Axiom which makes PA2 a second-order theory. Since Gödel's First Incompleteness Theorem applies to PA2, there is a sentence (which we call "$G2$") that is formally undecidable in PA2. In other words,

$$\text{PA2} \nvdash G2 \text{ and PA2} \nvdash \neg G2.$$

The Induction Axiom says that every object in the universe is a natural number, so $\mathfrak{N}$ is the only model of PA2. Either $G2$ is true in $\mathfrak{N}$, or $\neg G2$ is true in $\mathfrak{N}$. Since $\mathfrak{N}$ is the only model of PA2, one of these two sentences is true in every model of PA2, which means that one of them is valid in PA2. Yet they are both unprovable in PA2. We have shown that there is a second-order theory in which a valid sentence is not provable. Hence, the Completeness Theorem does not hold for second-order logic. This is not surprising since we saw in Section 5.5 that the Compactness Theorem, which is closely related to the Completeness Theorem, fails for second-order logic. To some logicians, the fact that second-order logic is not complete strengthens the argument that Mathematics should be based on first-order logic.

CHAPTER 7
CONSISTENCY OF ARITHMETIC

7.1 The Second Incompleteness Theorem

When Gödel made his unexpected announcement in September 1930, he made no mention of the second theorem that appeared in the paper he submitted for publication two months later (which appeared in print in 1931). A few days after submitting it, he received a letter from von Neumann. The letter said that following their discussion in September, he had made a remarkable discovery: *if an effective theory of arithmetic is consistent, then its consistency is unprovable in that theory.* But this was precisely the second theorem that Gödel had already included in his paper, and he had to break the news to a disappointed von Neumann who thought he might have been the first to make this important discovery.

The second theorem is known today as the *"Second Incompleteness Theorem"* (even though it has nothing to do with incompleteness). In his paper, Gödel gave only an outline of its proof. He may have intended give a complete proof in a subsequent paper, which would explain the "I" in the title of the one he submitted. But the second instalment of the paper never appeared. A complete proof of the Second Incompleteness Theorem did not appear until 1939 in the second volume of *Foundations of Mathematics*, a work jointly authored by Hilbert and his collaborator Paul Bernays.

The Second Incompleteness Theorem applies to *"Principia Mathematica and Related Systems"*. As we did for the First Theorem, we will treat it as though it applies just to PA. The second theorem then says that

> If PA is consistent, then its consistency cannot be proved in PA.   [7.1]

Although PA was designed to make and prove statements about numbers, it should not surprise you now that we are able to use $\mathcal{L}_{PA}$ to talk about proving the consistency of PA in PA. It is, of course, the device of Gödel numbering that extends the expressive power of $\mathcal{L}_{PA}$ so that it can express the concept of the consistency of PA.

Our first task is to find a sentence in $\mathcal{L}_{PA}$ (which we call *"ConPA"*) such that

> *ConPA* asserts that PA is consistent, or in other words,
>
> PA is consistent if and only if *ConPA* is true in $\mathfrak{N}$.

Axiom P1 of PA says that

$$\forall u\ (\ \neg\ s(u) = 0\ ).$$

Using the $\forall$-Elimination Rule, and substituting 0 for u, we conclude that

$$PA \vdash \neg\ s(0) = 0.$$

Now suppose that

$$PA \vdash s(0) = 0.$$

Then PA is inconsistent (because $PA \vdash s(0) = 0$ and $PA \vdash \neg s(0) = 0$).

Conversely, if PA is inconsistent, then every sentence is provable in PA, so

$$PA \vdash s(0) = 0.$$

Therefore,

PA is inconsistent if and only if $PA \vdash s(0) = 0$. [7.2]

Recall that for any sentence φ,

$PA \vdash \varphi$ if and only if $\exists x\ PrfPA(x, \ulcorner\varphi\urcorner)$ is true in $\mathfrak{N}$.

Now let "c" denote the Gödel number of the sentence $s(0) = 0$. Then

$PA \vdash s(0) = 0$ if and only if $\exists x\ PrfPA(x, \bar{c})$ is true in $\mathfrak{N}$. [7.3]

By combining [7.2] and [7.3], we conclude that

PA is inconsistent if and only if $\exists x\ PrfPA(x, \bar{c})$ is true in $\mathfrak{N}$,

or   PA is consistent if and only if $\neg \exists x\ PrfPA(x, \bar{c})$ is true in $\mathfrak{N}$. [7.4]

We name the sentence $\neg \exists x\ PrfPA(x, \bar{c})$ as "*ConPA*". Then [7.4] says that

PA is consistent if and only if *ConPA* is true in $\mathfrak{N}$,

or   *ConPA* expresses the informal statement that PA is consistent. [7.5]

We have found a sentence *ConPA* (which is a sentence in $\mathcal{L}_{PA}$) that expresses the consistency of PA. Proving *ConPA* in PA is "proving the consistency of PA in PA". Therefore, showing that *ConPA* is unprovable in PA would show that the consistency of PA cannot be proved in PA. Now we can re-state [7.1], the Second Incompleteness Theorem, as follows:

If PA is consistent, then $PA \nvdash ConPA$. [7.6]

If PA is inconsistent, then every sentence in $\mathcal{L}_{PA}$ (including *ConPA*) is provable in PA. Thus the Second Incompleteness Theorem ironically implies that

PA is consistent if and only if $PA \nvdash ConPA$,

or,   PA is consistent if and only if its consistency is unprovable in PA.

7.2 Proof of the Second Incompleteness Theorem

The proof of the Second Incompleteness Theorem is long and difficult, but like the First Theorem, the essential idea behind the proof is simple and elegant. The First Theorem is a metamathematical theorem informally proved in an informal language (originally in German). The Second Incompleteness Theorem is proved by showing that *part of the informal proof of the First Incompleteness Theorem can be written as a formal proof in* PA.

To prove the First Incompleteness Theorem, Gödel constructed a sentence G in $\mathcal{L}_{PA}$ such that G is the sentence $\neg \exists x \, PrfPA(x, \overline{\ulcorner G \urcorner})$, and proved that

if PA is consistent, then PA$\nvdash G$.   **[7.7]**

[7.7] is half of the First Incompleteness Theorem. (We do not need the other half that says if PA is ω-consistent, then PA$\nvdash \neg G$.)

In order to formally prove [7.7] in PA, it must first be expressed in $\mathcal{L}_{PA}$. We have already laid the groundwork for doing this. [7.5] and [6.27] in Section 6.7 are shown again below:

    ConPA expresses the informal statement that PA is consistent.   **[7.5]**

    G expresses the informal statement that PA$\nvdash G$.   **[6.27]**

Combining [7.5] and [6.27], we have the following:

    ConPA → G expresses the informal statement that
    if PA is consistent, then PA$\nvdash G$,

or   *ConPA* → G expresses the informal statement [7.7].   **[7.8]**

If you skipped Section 6.7, you would have missed [6.27]. But [6.19] in Section 6.6 says the same thing as [6.27]:

    G is true in $\mathfrak{N}$ if and only if G is not provable in PA.   **[6.19]**

Gödel proved [7.7] informally as part of the First Incompleteness Theorem. His proof was constructive and finitary, so consequently, it can be re-written as a formal proof in PA. According to [7.8], when an informal proof proves [7.7], the sentence proved by the formal version of the proof is *ConPA* → G.

Therefore, having a formal proof in PA of half of the First Incompleteness Theorem means that

    ConPA → G is provable in PA,

or   PA$\vdash$ *ConPA* → G.   **[7.9]**

We are now ready to "prove" the Second Incompleteness Theorem as expressed by [7.6]. In other words, we want to show this:

If PA is consistent, then PA⊬*ConPA*. [7.6]

To prove [7.6], we start by assuming that PA is consistent. Next, we want to show that PA⊬*ConPA*. To do that, we assume that

PA⊢*ConPA* [7.10]

and look for a contradiction. According to [7.9] and [7.10],

ConPA → *G* and *ConPA* are both provable in PA.

If we write the two proofs one after the other, we have a single proof in PA which includes the lines *ConPA* → *G* and *ConPA*. Applying the →-Elimination Rule to these two lines, we conclude that

PA⊢*G*.

But we have assumed that PA is consistent, and under this assumption, the First Incompleteness Theorem tells us that

PA⊬*G*.

Having shown that PA⊢*G* and PA⊬*G*, we have a contradiction. Our two assumptions that PA is consistent and PA⊢*ConPA* cannot both be true. Hence,

If PA is consistent, then PA⊬*ConPA*,

or If PA is consistent, then its consistency cannot be proved in PA (which implies it is unprovable in any theory weaker than PA).

This is the Second Incompleteness Theorem when applied to PA.

If T is any reasonable theory of arithmetic, then the First Incompleteness Theorem holds for T, and there is a sentence $ConT$ in $\mathcal{L}_T$ that expresses the consistency of T. It can be shown (in the way that we indicated for PA) that if T is consistent, then $ConT$ cannot be proved in T, or in any theory weaker than T. This conclusion applies to second-order theories of arithmetic, *Principia Mathematica*, and set theory.

7.3 Proving Consistency

If we wish to prove that some theory (such as PA or Grp) is consistent, that proof must itself be carried out in some system. The system that vouches for the consistency of some theory should be as reliable and trustworthy as possible. In Hilbert's words, the vouching process should be "finitary". For a simple theory such as Grp, a finitary vouching process indeed exists. The process consists of showing by a finitary process that the theory has a model, and we described how this could be done for Grp in Section 3.3.

But showing that a more complex theory such as PA has a model is not a finitary process. PA has infinitely many axioms, and the universe of any model is infinite. We need a more direct method if we want to show that PA is consistent. One way is to find a theory V (for "Vouching theory") that can prove PA is consistent. But if V is less trustworthy than PA (i.e. V is more likely to be inconsistent) then using V to prove the consistency of PA has little value. It would be like asking for a possibly corrupt politician to be investigated by a colleague with an even more dubious reputation.

The trustworthiness of a theory is related to its proving power. Suppose theory A is more powerful than theory B, which means that every theorem of B is a theorem of A, but not vice versa. Then it is not possible for B alone to be inconsistent because if B can prove a contradiction, so can A. However, it is possible for only A to be inconsistent. Therefore, A (the more powerful one) is less trustworthy than B. (Theories are like people; the more power they have, the less trustworthy they become.) To prove that PA is consistent, V needs to have a certain amount of power. But there is not much point using a theory V that is more powerful (and hence less trustworthy) than PA. We are caught in a squeeze between using a theory V with enough power, but not too much.

The Second Incompleteness Theorem leaves us very little wriggle room within this squeeze between too little power and too much. If we take V to be PA, the Second Theorem says that V would not have enough power. But increasing the power of this V would make it more powerful and less trustworthy than PA. We are in the following situation:

> if V is less powerful than PA (or has the same power as PA) then the Second Theorem says that V cannot prove the consistency of PA if PA is consistent,
>
> if V is more powerful than PA then a proof in V that PA is consistent has little value.

The situation appears hopeless until we realize that we can have two theories where neither is more powerful than the other. Each theory can have theorems that are not theorems of the other. Therefore, it might be possible to have a theory that can prove the consistency of PA without being more powerful (and less trustworthy) than PA.

What we need is a theory V such that
1. V can prove that PA is consistent (which implies that V is not less powerful than PA since V can prove something that PA cannot),
2. V is not more powerful than PA (which implies that V is not less trustworthy than PA, so a proof in V that PA is consistent has value).

In 1936, Gerhard Gentzen came up with such a theory, one that was substantially different from PA. In this system, he proved that PA is consistent, but his proof is not finitary. Even before Gentzen's theorem, there was reason to believe that PA is consistent. Confidence in the consistency of PA increased after he proved it in a theory that is *not less* trustworthy than PA. Our confidence would be increased even more if the consistency of PA could be proved in a theory that is *more* trustworthy than PA, but the Second Incompleteness Theorem shows that there is no such theory.

Both Incompleteness Theorems have historical significance because mathematicians expected that the axiomatic method would bring certainty and consistency back to Mathematics after a period of turmoil. The Second Incompleteness Theorem showed that the central problem of proving consistency cannot have a perfect solution. In hindsight, the expectations placed on the axiomatic method were much too optimistic, and the two Theorems forever expelled mathematicians from a Garden of Eden that they had imagined for themselves.

CHAPTER 8
MISCONCEPTIONS ABOUT INCOMPLETENESS

8.1 The Incompleteness Theorems and Philosophy

Although there is no dispute over what the two Incompleteness Theorems say mathematically, they have long had a mystique about them that has made them the subject of philosophical speculation. As a result, several misconceptions about the Incompleteness Theorems have arisen. We now list a few of the more egregious examples, and attempt to clarify them.

8.2 Some truths in $\mathfrak{N}$ are unprovable in any reasonable theory

Stated more precisely, this misconception says there is a sentence that is true in $\mathfrak{N}$, but unprovable in any reasonable theory of arithmetic. This misconception arises as a result of switching the order of informal quantifiers.

The First Incompleteness Theorem says that in any reasonable theory of arithmetic, there is a sentence that is true in $\mathfrak{N}$, but unprovable in that theory. The Theorem says that

    (In any reasonable theory) (There is a sentence such that)
    (The sentence is true and unprovable).   **[8.1]**

The misconception says that

    (There is a sentence such that) (In any reasonable theory)
    (The sentence is true and unprovable).   **[8.2]**

The only difference between [8.1] and [8.2] is the order of the informal quantifiers. Switching the order of quantifiers in a sentence changes its meaning. For example,

    Everyone has a mother

or   (For every Y) (There is an X such that) (X is the mother of Y)

does not have the same meaning as

    There is someone who is everyone's mother

or   (There is an X such that) (For every Y) (X is the mother of Y).

Switching the order of the quantifiers in the correct statement [8.1] turns it into the incorrect statement [8.2].

It is easy to see that the misconception is incorrect. Any true but unprovable sentence can be added as a new axiom to a reasonable theory of arithmetic, creating a new reasonable theory. In the new theory, that sentence is provable. In practice, the problem faced by mathematicians is deciding whether a formally undecidable sentence is true or not. In such cases, other "evidence" must be considered to make this decision. For example, with Gödel's G, there is evidence to show that it is true in $\mathfrak{N}$. We will see examples in set theory of formally undecidable sentences for which there is no conclusive evidence one way or the other.

8.3 Human intuition can prove more than axiomatic theories can

This misconception arises chiefly from the fact that PA cannot prove G, but we can recognize that G is true in $\mathfrak{N}$, and even prove it informally. The natural conclusion is that PA cannot match human intuition in establishing the truths of arithmetic. Furthermore, every reasonable theory of arithmetic has a sentence asserting its own unprovability that we know is true but unprovable. Thus, our intuition is stronger than many axiomatic theories.

In clarifying this misconception, we will restrict our attention to PA, but the conclusions we reach about PA apply to any reasonable theory of arithmetic. What makes this misconception tricky is that the argument given in the above paragraph is almost correct. What the argument neglects to mention is that we need to assume that PA is consistent before concluding that G is true and unprovable. Thus, our "knowledge" that G is true depends on our "knowing" that PA is consistent. Believers of the misconception reply that this is where humans excel over axiomatic theories; based on our intuition, we "know" that PA is consistent, but PA has no intuition and no way to establish the truth of G.

Now we are at the heart of the misconception. An axiomatic theory "knows" only what can be deduced from its axioms, and we are the ones who decide what its axioms are. Before we can claim to "know" that G is true and unprovable, we must first claim to "know" that PA is consistent. Yet we did not give this vital "knowledge" to PA; we did not make *ConPA* an axiom. After withholding "knowledge" from PA that we allowed ourselves to use, it is not surprising that we "know more" than PA. We are not competing on a level playing field.

If we add *ConPA* as an additional axiom to PA, we would have a new theory. We will call it "PA+ConPA". In Section 7.2, we saw that

$$\text{PA} \vdash ConPA \rightarrow G. \qquad [7.9]$$

Every theorem of PA is also a theorem of PA+ConPA because every proof in PA is also a proof in PA+ConPA. Therefore,

$$\text{PA+ConPA} \vdash ConPA \rightarrow G. \qquad [8.3]$$

Since *ConPA* is an axiom of PA+ConPA,

$$\text{PA+ConPA} \vdash ConPA. \qquad [8.4]$$

Combining the two proofs in [8.3] and [8.4] into a single proof, and applying the $\rightarrow$-Elimination Rule, we conclude that

$$\text{PA+ConPA} \vdash G.$$

Thus, if we strengthen PA by granting it the "knowledge" that PA is consistent, we have a new theory that can prove G. With a level playing field, axiomatic theories can "know" just as much as we do.

Although PA+ConPA can prove G, it cannot prove its own version of G (a sentence in $\mathcal{L}_{PA}$ that asserts its own unprovability in PA+ConPA) if it is consistent. Furthermore, while PA+ConPA can prove the consistency of PA, it cannot prove its own consistency if it is consistent. We cannot escape the grasp of the Incompleteness Theorems.

The contention that human intuition can prove more than axiomatic theories can has been supported by some authors on more speculative grounds. Some claim that we have consciousness while theories do not, which may well be true. Some claim that our brains have extra powers due to the strange effects of quantum physics, which may also be true. Some have even claimed that the Incompleteness Theorems provide evidence for the existence of God. But there have not been any firm conclusions drawn from such speculation. When thinking about ideas like these, bear in mind exactly what the Incompleteness Theorems say; that's what we know for sure.

8.4   The consistency of Peano Arithmetic can never be proved

This topic was discussed in Chapter 7, and we now repeat some of the points we made. Whether or not the consistency of PA can be proved depends on what we mean by "proved". Three possible meanings are described below.

If "proved" means provable in PA or in a theory *more* trustworthy than PA, then indeed PA can never be proved to be consistent. This is what the Second Incompleteness Theorem tells us.

If "proved" means provable in a theory that is *not less* trustworthy than PA, then Gentzen's proof shows that it is possible to prove the consistency of PA. Gentzen's system is neither more nor less trustworthy than PA because it is neither stronger nor weaker than PA in proving theorems.

If "proved" means provable in ZFC (set theory), then again, it is possible to prove the consistency of PA. ZFC can prove that the natural numbers embedded in every one of its models satisfy the axioms of PA, proving that PA has a model. (Thus, *ConPA* is an example of a sentence in $\mathcal{L}_{PA}$ that is unprovable in PA, but provable in ZFC after being translated to a sentence in $\mathcal{L}_{ZF}$.) Provability in ZFC is the standard that is usually accepted in other branches of Mathematics when settling a question. But in this case, there is some reluctance to accept the judgement of the "Supreme Court". The reason is simple. ZFC is more powerful and hence less trustworthy than PA, making the blessing of ZFC logically pointless. But is this "fair" to PA?

Let's consider the parallel situation when a real court passes judgement on a person. The Supreme Court has ruled that Jane Doe is honest. But everyone knows that Jane is more honest than the Court itself. This makes the Court's ruling pointless; her reputation is not enhanced by a favourable judgement from such a court, and we thus discount the value of this judgement. But we accept favourable judgements from this Court when they are bestowed upon people who are less trustworthy than Jane. Jane's excellent reputation is the reason that she cannot benefit from a favourable Supreme Court ruling. If we still doubt that she is honest, then we must admit that we are holding her to a higher standard than other people. Is this "fair" to Jane?

To some mathematicians, PA *should* be held to a higher standard than other theories because arithmetic is such a fundamental component of Mathematics. According to Kronecker, "*Natural numbers were created by God, everything else is the work of men.*" Gentzen's system is more trustworthy than ZFC, so his proof indeed holds PA to a higher standard.

CHAPTER 9
SET THEORY

9.1 Cantor's Set Theory

In 1938, Gödel partially answered two important questions in set theory. To understand them, we need to examine some ideas of Cantor, who first studied sets mathematically in the 1870s. At that time, there was no apparent need for axioms about sets. The idea of a set was a fundamental one; sets were simply "out there" in the real world, ready for us to study. Number theorists had studied numbers with the attitude that everyone knows what numbers are, and Cantor approached sets in the same way. The ideas that Cantor discovered are as fundamental to sets as addition and multiplication are to numbers. Cantor's work made set theory (without axioms) into a branch of Mathematics.

9.2 Basic Concepts in Set Theory

In the 19th century, any collection of objects could be considered as a "set", no matter how many objects are involved. This was the carefree approach of all set theorists including Cantor and Frege. After the discovery of Russell's Paradox in 1901, set theorists were forced to take a more cautious approach, leading to the formulation of axiomatic set theory. Although Russell's Paradox derailed Frege's ambitious program, it did little damage to Cantor's work, which retained a central role in axiomatic set theory. We next look at the concepts introduced by Cantor. First, some very basic concepts and notation.

The objects that belong to a set are called its *"members"* or *"elements"*. For example, let "$\mathbb{E}$" denote the set of even natural numbers. Then 6 is a member of $\mathbb{E}$, but 7 is not. The informal symbol $\in$ means "is a member of", so we have

$6 \in \mathbb{E}$, $7 \notin \mathbb{E}$.

Two sets are considered to be equal (i.e. they are the *same* set) if they have exactly the same members. When there is no ambiguity, we can list the members, or describe them within curly brackets. Thus,

$\mathbb{E} = \{0, 2, 4, 6, 8, 10, \ldots, 2 \times n, \ldots\}$

$\mathbb{E} = \{n \in \mathbb{N} : n \text{ is even}\}$. ($\mathbb{N}$ is the set of all natural numbers).

Repeating a member or re-ordering the list does not change the set. Thus,

$\{2, 4, 5, 7\} = \{7, 4, 2+3, 7, 3+1, 2\}$.

We say a set is *empty* if it has no members. Any two empty sets have exactly the same members, so they are the same set. Therefore, we are justified in calling it "*the* empty set". It is usually denoted by "∅". Thus,

$$\emptyset = \{\ \}.$$

Note that ∅ is not "nothing". It is the set with no members. Note that {∅} is a set with one member, that member being the empty set.

When $\mathbb{X}$ and $\mathbb{Y}$ are sets, we say that $\mathbb{X}$ is a *subset* of $\mathbb{Y}$ (denoted by "$\mathbb{X} \subseteq \mathbb{Y}$") if every member of $\mathbb{X}$ is a member of $\mathbb{Y}$. Note that for any set $\mathbb{X}$,

$$\emptyset \subseteq \mathbb{X} \text{ and } \mathbb{X} \subseteq \mathbb{X}.$$

The *union* of $\mathbb{X}$ and $\mathbb{Y}$ (denoted by "$\mathbb{X} \cup \mathbb{Y}$") is the set whose members are all the members of $\mathbb{X}$ and all the members of $\mathbb{Y}$ (and nothing else).

Two sets are said to be *disjoint* if they have no members in common. In other words, no member of either set is a member of the other set. For example,

$\mathbb{E}$ and $\{5, 7, 13, \mathbb{E}\}$ are disjoint.

The second set above has four members (three numbers and one set), and none of them is a member of the first set because none of them is an even number.

9.3 Equipotent Sets

For a set with a finite number of members, it is easy to define its "size"; it is the number of members that it has. Cantor wanted to extend the concept of size to include all sets, including infinite ones such as $\mathbb{N}$. Obviously, "number of members" does not work as a definition of "size" for infinite sets. Instead of worrying about what size *is*, Cantor defined a new criterion for what it means for two sets to have the "*same size*", a criterion that applies to all sets. (In a similar way, early physicists knew what it meant for two objects to have the *same temperature* without understanding what temperature *is*.) We use the word "equipotent" to mean "same size" according to Cantor's criterion.

Two sets $\mathbb{X}$ and $\mathbb{Y}$ are said to be "*equipotent*" if

> it is *possible* for members of $\mathbb{X}$ to be matched up one-to-one with members of $\mathbb{Y}$, forming pairs with no member from either set left unmatched. [9.1]

It is easily seen that two finite sets are equipotent if and only if they have the same number of members. Thus, for finite sets, the idea of being equipotent is consistent with the idea of having the same size.

For an example with infinite sets, $\mathbb{N}$ (the set of natural numbers) and $\mathbb{E}$ (the set of even natural numbers) are equipotent. Here is a match-up that confirms it:

Every n in $\mathbb{N}$ is matched with $2 \times n$ (which is a member of $\mathbb{E}$). **[9.2]**

With this match-up, every member in each set is paired with one and only one partner from the other set, and no member in either set is left without a partner.

With infinite sets that are equipotent, we can make "careless" match-ups that have left-over members. Consider the two match-ups described below:

Every n in $\mathbb{N}$ is matched with $2+2 \times n$ (which is a member of $\mathbb{E}$), **[9.3]**

and Every n in $\mathbb{E}$ is matched with n (which is a member of $\mathbb{N}$). **[9.4]**

Both these match-ups are one-to-one partnerships, but they have left-overs (unmatched members). In [9.3], 0 in $\mathbb{E}$ is left unmatched. In [9.4], every odd number in $\mathbb{N}$ is left unmatched. The fact that there are one-to-one match-ups that have left-overs is irrelevant. [9.2] is a one-to-one match-up between the members of $\mathbb{N}$ and $\mathbb{E}$ that leaves no unmatched members, and that is all we need to declare that $\mathbb{N}$ and $\mathbb{E}$ are equipotent.

At this point, some mathematicians objected to what Cantor was doing. $\mathbb{E}$ is the set we have after removing many members from $\mathbb{N}$. "Equipotent" is supposed to generalize the notion of "same size". So how can $\mathbb{N}$ be the same size as $\mathbb{E}$? This surely shows the folly of treating infinite sets as entities that can be compared. Cantor's position was that he had simply made a definition, and by his definition, $\mathbb{N}$ and $\mathbb{E}$ are equipotent.

We abbreviate "$\mathbb{X}$ and $\mathbb{Y}$ are equipotent" by "$\mathbb{X} \equiv \mathbb{Y}$". $\not\equiv$ means not equipotent. If $\mathbb{W}$ is finite, let "$|\mathbb{W}|$" denote the number of members in $\mathbb{W}$, or the size of $\mathbb{W}$.

For all finite sets $\mathbb{W}$, $\mathbb{X}$, and $\mathbb{Y}$, the relationship of having the same size has some obvious properties which are just facts about equality between numbers:

$|\mathbb{W}| = |\mathbb{W}|$

If $|\mathbb{W}| = |\mathbb{X}|$ and $|\mathbb{X}| = |\mathbb{Y}|$ then $|\mathbb{W}| = |\mathbb{Y}|$.

It is easy to verify (and reassuring to know) that the relationship of being equipotent has the same properties. For all sets $\mathbb{W}$, $\mathbb{X}$, and $\mathbb{Y}$,

$\mathbb{W} \equiv \mathbb{W}$

If $\mathbb{W} \equiv \mathbb{X}$ and $\mathbb{X} \equiv \mathbb{Y}$ then $\mathbb{W} \equiv \mathbb{Y}$.

These properties are just facts about one-to-one match-ups without left-overs.

Having checked that the properties of having the same size when referring to finite sets are also properties of being equipotent, we can now be confident that equipotency is a proper generalization of "same size in finite sets".

9.4 Comparing Potencies

For finite sets $\mathbb{X}$ and $\mathbb{Y}$, we know what it means for $\mathbb{X}$ to be "smaller than or have the same size as" $\mathbb{Y}$, or to be "not bigger than" $\mathbb{Y}$. It means that

$$|\mathbb{X}| \leq |\mathbb{Y}|. \qquad [9.5]$$

To generalize this notion of "not bigger than" to apply to all sets, we express [9.5] without mentioning size, but in terms of equipotency. For finite sets $\mathbb{X}$ and $\mathbb{Y}$, [9.5] is true if and only if

$\mathbb{X}$ is equipotent to a subset of $\mathbb{Y}$. $\qquad [9.6]$

We take [9.6] as the generalization of [9.5] that applies to all sets. We write [9.6] symbolically as

$$\mathbb{X} \preccurlyeq \mathbb{Y}.$$

We need to check that this is a proper generalization of [9.5], which means checking that the essential properties of the relationship [9.5] are also properties of [9.6]. Here are some important properties of [9.5].

For all finite sets $\mathbb{W}$, $\mathbb{X}$, and $\mathbb{Y}$,

$|\mathbb{W}| \leq |\mathbb{W}|$

If $|\mathbb{W}| \leq |\mathbb{X}|$ and $|\mathbb{X}| \leq |\mathbb{Y}|$ then $|\mathbb{W}| \leq |\mathbb{Y}|$

If $|\mathbb{W}| \leq |\mathbb{X}|$ and $|\mathbb{X}| \leq |\mathbb{W}|$ then $|\mathbb{W}| = |\mathbb{X}|$.

The above three properties are also properties of $\preccurlyeq$.

For all sets $\mathbb{W}$, $\mathbb{X}$, and $\mathbb{Y}$,

$\mathbb{W} \preccurlyeq \mathbb{W}$

If $\mathbb{W} \preccurlyeq \mathbb{X}$ and $\mathbb{X} \preccurlyeq \mathbb{Y}$ then $\mathbb{W} \preccurlyeq \mathbb{Y}$.

If $\mathbb{W} \preccurlyeq \mathbb{X}$ and $\mathbb{X} \preccurlyeq \mathbb{W}$ then $\mathbb{W} \equiv \mathbb{X}$. $\qquad [9.7]$

If $\preccurlyeq$ did not have these three properties, we would not be justified in considering it as a generalization of "not bigger than" for finite sets.

The third property shown above ([9.7]) is known as the "Cantor-Bernstein Theorem"; it was first stated by Cantor without proof, and then proved by Felix Bernstein. The proof is not very easy.

SET THEORY

There is one more relationship between finite sets that we wish to generalize to all sets. For finite sets $\mathbb{X}$ and $\mathbb{Y}$, $\mathbb{X}$ is smaller than $\mathbb{Y}$ when

$|\mathbb{X}| < |\mathbb{Y}|$, or equivalently,

$|\mathbb{X}| \leq |\mathbb{Y}|$ and $|\mathbb{X}| \neq |\mathbb{Y}|$.

Here then is the generalization of "smaller than" for all sets $\mathbb{X}$ and $\mathbb{Y}$:

$\mathbb{X} \preccurlyeq \mathbb{Y}$ and $\mathbb{X} \not\equiv \mathbb{Y}$.

We abbreviate this relationship as "$\mathbb{X} \prec \mathbb{Y}$". The obvious properties of the "smaller than" relationship for finite sets are also properties of $\prec$. For example,

If $\mathbb{W} \prec \mathbb{X}$ and $\mathbb{X} \prec \mathbb{Y}$ then $\mathbb{W} \prec \mathbb{Y}$, and

If $\mathbb{W} \prec \mathbb{X}$ and $\mathbb{X} \equiv \mathbb{Y}$ then $\mathbb{W} \prec \mathbb{Y}$.

Even for infinite sets, we will say that "$\mathbb{X}$ is smaller than $\mathbb{Y}$" when $\mathbb{X} \prec \mathbb{Y}$.

As the concept of size for finite sets was successfully generalized to apply to all sets, the initial scepticism regarding the study of infinite sets began to fade. For example, the fact that we can remove some members from a set without changing its "size" was no longer seen as a reason to reject Cantor's theory; it is merely a property possessed by infinite sets.

9.5 The Power Set of a Set

For any set $\mathbb{X}$, the set of all its subsets is called the "*power set*" of $\mathbb{X}$ and denoted by "$\mathcal{P}(\mathbb{X})$". It is easier to say "the power set of" than "the set of all subsets of". For example,

if $\mathbb{X} = \{5, 6, 7\}$, then

$\mathcal{P}(\mathbb{X}) = \{\, \emptyset, \{5\}, \{6\}, \{7\}, \{5, 6\}, \{5, 7\}, \{6, 7\}, \mathbb{X} \,\}$.

$\mathcal{P}(\mathbb{X})$ is a set with 8 members, and they are all subsets of $\mathbb{X}$.

Cantor proved the following important theorem on power sets:

For any set $\mathbb{X}$, $\mathbb{X} \prec \mathcal{P}(\mathbb{X})$.

In other words, any set is smaller than its power set. A set is never equipotent with its power set, but it is always equipotent to some subset of its power set.

To show that $\mathbb{X} \prec \mathcal{P}(\mathbb{X})$ for any set $\mathbb{X}$, we need to complete two steps:

Step 1. Show that $\mathbb{X}$ and some subset of $\mathcal{P}(\mathbb{X})$ are equipotent,

Step 2. Show that $\mathbb{X}$ and $\mathcal{P}(\mathbb{X})$ are not equipotent.

For Step 1, we let $\mathbb{T}$ be the set of all the subsets of $\mathbb{X}$ that have one member. Then $\mathbb{T}$ is a subset of $\mathcal{P}(\mathbb{X})$. We match up each $x \in \mathbb{X}$ with $\{x\} \in \mathbb{T}$. This match-up has no left-overs, so $\mathbb{X}$ and $\mathbb{T}$ are equipotent. For example, if $\mathbb{X} = \{5, 6, 7\}$, then $\mathbb{T} = \{\, \{5\}, \{6\}, \{7\} \,\}$ (which is a subset of $\mathcal{P}(\mathbb{X})$), and we would match 5 with $\{5\}$, 6 with $\{6\}$, and 7 with $\{7\}$.

For Step 2, we assume we have a one-to-one match-up between the members of $\mathbb{X}$ and $\mathcal{P}(\mathbb{X})$. For every $x \in \mathbb{X}$, we let "$\mathbb{S}_x$" be the name of the set in $\mathcal{P}(\mathbb{X})$ that has been matched with x. Each $\mathbb{S}_x$ is a subset of $\mathbb{X}$. We show that $\mathcal{P}(\mathbb{X})$ must have some member $\mathbb{Y}$ that has been left out of the match-up. Consider this set:

$$\mathbb{Y} = \{\, x \in \mathbb{X} : x \notin \mathbb{S}_x \,\}.$$

$\mathbb{Y}$ is a subset of $\mathbb{X}$ because all its members are members of $\mathbb{X}$. Thus $\mathbb{Y} \in \mathcal{P}(\mathbb{X})$. Any member x in $\mathbb{X}$ is a member of $\mathbb{Y}$ if and only if it is not a member of $\mathbb{S}_x$. Therefore, for any x, $\mathbb{S}_x$ is not $\mathbb{Y}$ because $\mathbb{S}_x$ and $\mathbb{Y}$ disagree on whether or not x is a member. But if $\mathbb{Y}$ had been matched up with some member z of $\mathbb{X}$, it would be equal to $\mathbb{S}_z$. Therefore, $\mathbb{Y}$ has not been matched with any member of $\mathbb{X}$. No matter what match-up we make between the members of $\mathbb{X}$ and the members of $\mathcal{P}(\mathbb{X})$, we can always find such a $\mathbb{Y}$. We have shown that is impossible to match the members of $\mathbb{X}$ with the members of $\mathcal{P}(\mathbb{X})$ without having at least one member in $\mathcal{P}(\mathbb{X})$ that is unmatched. Therefore, $\mathbb{X}$ and $\mathcal{P}(\mathbb{X})$ are not equipotent.

Combining Steps 1 and 2, we conclude that for any set $\mathbb{X}$, $\mathbb{X} \prec \mathcal{P}(\mathbb{X})$.

The above proof of Step 2 may be clarified by an example. Suppose $\mathbb{X}$ is $\mathbb{N}$, and we have a match-up between the members of $\mathbb{N}$ and the members of $\mathcal{P}(\mathbb{N})$. We want to show that some member of $\mathcal{P}(\mathbb{N})$ did not get matched up. Suppose the first few members of $\mathbb{N}$ are matched with members of $\mathcal{P}(\mathbb{N})$ as shown below.

- 0 is matched with $\{n \in \mathbb{N} : n$ is even$\}$ (we call this set "$\mathbb{S}_0$"),
- 1 is matched with $\{n \in \mathbb{N} : n$ is odd$\}$ (we call this set "$\mathbb{S}_1$"),
- 2 is matched with $\emptyset$ (we call this set "$\mathbb{S}_2$"),
- 3 is matched with $\{0, 2, 3\}$ (we call this set "$\mathbb{S}_3$"),
- 4 is matched with $\{1, 2, 5, 8\}$ (we call this set "$\mathbb{S}_4$").

Then we build the unmatched set $\mathbb{Y}$ in the following way:

- $0 \in \mathbb{S}_0$, so $0 \notin \mathbb{Y}$, making $\mathbb{S}_0$ and $\mathbb{Y}$ different,
- $1 \in \mathbb{S}_1$, so $1 \notin \mathbb{Y}$, making $\mathbb{S}_1$ and $\mathbb{Y}$ different,
- $2 \notin \mathbb{S}_2$, so $2 \in \mathbb{Y}$, making $\mathbb{S}_2$ and $\mathbb{Y}$ different,
- $3 \in \mathbb{S}_3$, so $3 \notin \mathbb{Y}$, making $\mathbb{S}_3$ and $\mathbb{Y}$ different,
- $4 \notin \mathbb{S}_4$, so $4 \in \mathbb{Y}$, making $\mathbb{S}_4$ and $\mathbb{Y}$ different.

SET THEORY

We do this for every n in $\mathbb{N}$, making n a member of $\mathbb{Y}$ if and only if n is not a member of $\mathbb{S}_n$ (the set matched with n). This makes $\mathbb{Y}$ and $\mathbb{S}_n$ different for every n. Thus, $\mathbb{Y}$ is a member of $\mathcal{P}(\mathbb{N})$ that is not matched with any member of $\mathbb{N}$.

In Section 6.11, we saw examples of diagonalization. Cantor's proof was the original one to use this technique. Staying with the above example of $\mathbb{S}_0$ to $\mathbb{S}_4$, the table below shows which numbers less than 6 are "in" or "out" of which set.

		0	1	2	3	4	5
$\{n : n \text{ is even}\} =$	$\mathbb{S}_0$	<u>in</u>	out	in	out	in	out
$\{n : n \text{ is odd}\} =$	$\mathbb{S}_1$	out	<u>in</u>	out	in	out	in
$\emptyset =$	$\mathbb{S}_2$	out	out	<u>out</u>	out	out	out
$\{0, 2, 3\} =$	$\mathbb{S}_3$	in	out	in	<u>in</u>	out	out
$\{1, 2, 5, 8\} =$	$\mathbb{S}_4$	out	in	in	out	<u>out</u>	in

For example, "in" appears in the $\mathbb{S}_1$ row under 3, meaning that 3 is "in" $\mathbb{S}_1$. To define the set $\mathbb{Y}$, we go down the diagonal (where "in" and "out" are underlined), replacing "in" by "out" and "out" by "in". For example, 4 is "out" of $\mathbb{S}_4$, which means 4 is "in" $\mathbb{Y}$. Thus, you can "see" that $\mathbb{Y}$ is different from every $\mathbb{S}_n$.

9.6 The Continuum Hypothesis

The real numbers are those that can be expressed with a decimal point with possibly infinitely many digits after the point. The set of real numbers (denoted by "$\mathbb{R}$") is also known as the "continuum" because when the real numbers are marked on a line, they fill it up continuously without any gaps.

Cantor proved that

$\mathbb{N} \prec \mathbb{R}$, or in other words, $\mathbb{N}$ is smaller than $\mathbb{R}$.

The proof is very similar to his "diagonal" proof that $\mathbb{N} \prec \mathcal{P}(\mathbb{N})$, but with a few more messy details to be checked. Having proved that $\mathbb{N}$ is smaller than $\mathbb{R}$, he asked if $\mathbb{R}$ has a subset that is bigger than $\mathbb{N}$, but smaller than $\mathbb{R}$. In other words, he asked if [9.8] below is true.

$\mathbb{R}$ has a subset $\mathbb{X}$ such that $\mathbb{N} \prec \mathbb{X} \prec \mathbb{R}$ [9.8]

He believed that the answer was "No", and this belief is called the "*Continuum Hypothesis*". It says that [9.8] is false, so that any set bigger than $\mathbb{N}$ cannot be smaller than $\mathbb{R}$ (the continuum). In other words, although $\mathbb{R}$ is bigger than $\mathbb{N}$, it is not "much" bigger. But neither Cantor nor anyone else could prove the Continuum Hypothesis. Neither could anyone show that it is false by finding a set $\mathbb{X}$ that satisfies [9.8]. So is it true or false? The question remained unanswered. $\mathbb{N}$ and $\mathbb{R}$ are the sets underpinning arithmetic and analysis, two major branches of Mathematics. It was natural to seek more knowledge about the relative "sizes" of these sets. In 1900, Hilbert, considered the Continuum Hypothesis to be so significant that settling it one way or another was the first on his list of 23 important unsolved problems.

Using the Cantor-Bernstein Theorem ([9.7]), it is not too difficult to prove that
$$\mathbb{R} \equiv \mathcal{P}(\mathbb{N}).$$
Consider the statement [9.9] below:

There is no set $\mathbb{X}$ such that $\mathbb{N} \prec \mathbb{X} \prec \mathcal{P}(\mathbb{N})$. [9.9]

Since $\mathbb{R} \equiv \mathcal{P}(\mathbb{N})$, [9.9] is true if and only if [9.8] is false. So [9.9] is another way of expressing the Continuum Hypothesis.

9.7 The Axiom of Choice

We defined the relationship $\mathbb{X} \preccurlyeq \mathbb{Y}$ for any two sets as

$\mathbb{X}$ is equipotent to a subset of $\mathbb{Y}$.

It generalizes the relationship
$$|\mathbb{X}| \leq |\mathbb{Y}|$$
for finite sets to apply to all sets. We then listed three essential properties of the "$|\mathbb{X}| \leq |\mathbb{Y}|$" relationship that are also possessed by the "$\mathbb{X} \preccurlyeq \mathbb{Y}$" relationship, making $\preccurlyeq$ an acceptable generalization of $\leq$. But there is one more property that we did not consider:

For all finite sets $\mathbb{X}$ and $\mathbb{Y}$, either $|\mathbb{X}| \leq |\mathbb{Y}|$ or $|\mathbb{Y}| \leq |\mathbb{X}|$.

(Note that "or" always means that both possibilities may be true.) The generalization of this fact to the $\preccurlyeq$ relationship is the following assertion:

For any sets $\mathbb{X}$ and $\mathbb{Y}$, either $\mathbb{X} \preccurlyeq \mathbb{Y}$ or $\mathbb{Y} \preccurlyeq \mathbb{X}$. [9.10]

[9.10] says that for any two sets, at least one of them is equipotent to a subset of the other. This seems intuitively "obvious"; just start pairing up the members of $\mathbb{X}$ and $\mathbb{Y}$ until one set has been completely paired up. For example if every member of $\mathbb{Y}$ has been paired with a member of $\mathbb{X}$, then $\mathbb{Y} \preccurlyeq \mathbb{X}$. So [9.10] "should" be provable, and Cantor tried to prove it. At that time, axioms for set theory had not yet been formulated, so "provable" meant "informally provable", not "provable in a formal theory using axioms and rules of inference". With informal proofs, there was a toolbox of informal "axioms and rules" that may be used in making a proof. But nobody could prove [9.10] using only the tools in the standard toolbox.

In Section 2.2, we mentioned that some important theorems could not be proved without making a controversial assumption called the "Axiom of Choice". One example is [9.10]; by assuming the Axiom of Choice as an extra tool, Cantor was able to informally prove [9.10]. This laid the groundwork needed to define the cardinal numbers, as explained in Appendix C.

It's about time we stated the Axiom of Choice. [9.11] below is one of many ways of expressing the *Axiom of Choice*.

> If $\mathbb{X}$ is any set of non-empty sets such that any two members of $\mathbb{X}$ are disjoint (i.e. have no members in common), then there exists a set that contains exactly one member from each member of $\mathbb{X}$. **[9.11]**

For example, suppose $\mathbb{X}$ is the following set of three sets:

$\mathbb{X} = \{\, \mathbb{E}, \{3, 5, 9\}, \{n \in \mathbb{N} : n \text{ is odd and } n > 9\} \,\}$.

($\mathbb{E}$ is the set of even natural numbers.) None of the members of $\mathbb{X}$ are empty, and any two of them are disjoint. The Axiom of Choice says there exists a set consisting of exactly one member from each set in $\mathbb{X}$. Such a set is called a "*choice set*" for $\mathbb{X}$. Here is a choice set for the $\mathbb{X}$ defined above: $\{14, 5, 17\}$. In this example, $\mathbb{X}$ consists of only three sets, and we didn't need the Axiom of Choice. We simply made three choices to show that a choice set exists. But when $\mathbb{X}$ is enormous (e.g. it consists of one set for each real number) and the sets in $\mathbb{X}$ are complex, it is less obvious that a choice set must exist.

The following argument suggests that the Axiom of Choice is true in the "real world" of sets. Just go through the entire set $\mathbb{X}$, and choose one member from each member of $\mathbb{X}$. Making such a choice is always possible because $\mathbb{X}$ consists of non-empty sets. Then we define $\mathbb{Y}$ to be the set of all the chosen members. We now have a set $\mathbb{Y}$ that consists of exactly one member from each member of $\mathbb{X}$. In other words, $\mathbb{Y}$ is a choice set for $\mathbb{X}$. If you find this argument convincing, then you believe the Axiom of Choice is true, and you are not alone. If the argument leaves you unconvinced, then you have doubts about the Axiom of Choice, and again, you are not alone.

9.8 Axiomatic Set Theory

When Zermelo formulated the Axiom of Choice in 1904, it was not an axiom in a formal theory in the way that P1 to P7 are formal axioms of PA. Set theory was not yet a formal theory, and the Axiom of Choice was meant as another tool to be used in informal proofs. Indeed, it was an essential tool if [9.10] was going to be a theorem. But as we have already mentioned, not everyone considered it to be a legitimate tool.

With uncertainty over the Continuum Hypothesis, controversy over the Axiom of Choice, and the more fundamental problem of contradictions, it was evident that set theory could no longer be studied using intuition alone; an axiomatic theory was needed. With suitable axioms, it was hoped that contradictions could be avoided, and progress could be made on unsolved problems. Axiomatising set theory does not mean that proofs would become formal. In practice, proofs would still be informal, but any questionable deduction could now be checked against the axioms and rules of inference. However, the task of finding suitable axioms was not easy. There were two conflicting needs:

Need 1. The usefulness and power of the concept of a "set" depend on its generality; any conceivable collection "should" be a set. Therefore, the axioms need to allow as many sets as possible to exist.

Need 2. Allowing unlimited freedom to declare any imaginable collection (e.g. "the set of all sets") as a set leads to sets that are too "big", and results in contradictions. Therefore, there is a need to restrict what objects may be considered as "sets".

For example, accepting the Axiom of Choice respects Need 1 because we can certainly imagine that every set of non-empty disjoint sets has a choice set. But the Axiom does not specify how the members of a choice set are chosen. That raised concerns that it does not pay enough heed to Need 2; it might be too liberal, allowing too many sets, and perhaps lead to a contradiction.

In the 1920s, a set of first-order axioms was produced, creating Zermelo-Fraenkel set theory, or ZF. Its language $\mathcal{L}_{ZF}$ includes the Equality predicate symbol = and one non-logical predicate symbol ε. The objects in the universe of any model of ZF are called "sets". If $\mathbb{X}$ and $\mathbb{Y}$ are sets in the universe of some model of ZF, and $\mathbb{X} \, \varepsilon \, \mathbb{Y}$ is true in that model, we say that $\mathbb{X}$ is a member of $\mathbb{Y}$, or $\mathbb{Y}$ contains $\mathbb{X}$. Axioms of ZF ensure that ε in $\mathcal{L}_{ZF}$ captures the meaning of the predicate "__ is a member of __" (which we abbreviated as "∈"), just as axioms of PA ensure that + in $\mathcal{L}_{PA}$ captures the meaning of plus. The axioms grant certain entities the right to be a set (i.e. to be in the universe). For example, one axiom implies that if $\mathbb{X}$ and $\mathbb{Y}$ are sets, there exists a set whose members are those of $\mathbb{X}$ and $\mathbb{Y}$ combined. In other words, the union of two sets is a set. By modestly meeting the minimum demands of Need 1, it was expected that ZF was consistent. Thus, ZF appeared to strike the right balance between the two needs. See Appendix B for more on the Zermelo-Fraenkel axioms.

Recall that although PA was designed to characterize the natural numbers, it has unintended models. Like PA, ZF also has different models. But the notion of the "real world of sets" is difficult to pin down, so unlike PA, there is no "intended model" or "standard model" of ZF.

There is a sentence in $\mathcal{L}_{ZF}$ that expresses the Axiom of Choice (expressed informally by [9.11]). We call that sentence "*AC*". It was not included as one of the axioms of ZF because it was not universally accepted. But it was indispensable for proving many desirable theorems, so another version of set theory called "ZFC" was created by adding *AC* to the axioms of ZF. With two different set theories (one stronger than the other), the phrase "theorem of set theory" is now ambiguous. Whether or not the Axiom of Choice is used in a proof is sometimes significant, so it is useful to know if a theorem can be proved in ZF (and ZFC), or only in ZFC. For example, [9.10] (or more precisely, the sentence in $\mathcal{L}_{ZF}$ that expresses [9.10]) is provable in ZFC, but not in ZF.

There is a sentence in $\mathcal{L}_{ZF}$ that expresses the Continuum Hypothesis (expressed informally by [9.9]). We call that sentence "*CH*". Neither *CH* nor ¬*CH* are obviously true, so neither sentence was included among the axioms of ZF.

Merely having all this formal machinery in place does not answer any of the questions about the Axiom of Choice or the Continuum Hypothesis. But those problems are now well-defined, and could be approached systematically. For the Continuum Hypothesis we can investigate if *CH* is provable in ZF, or in ZFC. If not, can ¬*CH* be proved? Answering these questions may settle the problem. The path to resolving the Axiom of Choice was also clearly laid out. Those who think *AC* is false would hope that ¬*AC* can be proved in ZF. Those who merely have doubts about *AC* would expect that ZF cannot prove *AC*. Those who believe *AC* to be true would hope that it can be proved in ZF.

9.9 Gödel's Partial Solution

In 1938, Gödel proved two metamathematical theorems that partially solved the problems of the Axiom of Choice and the Continuum Hypothesis.

For the Axiom of Choice, Gödel proved that

If ZF is consistent, so is ZFC. [9.12]

This is an example of a *relative consistency* theorem, stating that one theory is consistent if another one is. Now suppose $ZF \vdash \neg AC$. Then $ZFC \vdash \neg AC$, and since $ZFC \vdash AC$, ZFC is inconsistent. Therefore, [9.12] implies that

If ZF is consistent, then $ZF \nvdash \neg AC$. [9.13]

[9.13] tells us that it is futile to try proving in ZF that the Axiom of Choice is false. Another consequence of [9.12] is more apparent if we state it in reverse:

If ZFC is inconsistent, so is ZF.

Those who need the Axiom of Choice in a proof may now safely assume it, secure in the knowledge that the assumption is not the cause of any inconsistency. If ZFC is inconsistent, the fault lies with ZF, not with AC.

For the Continuum Hypothesis, we let "ZFC+CH" denote the theory ZFC with CH added as an additional axiom. Gödel proved that

If ZF is consistent, so is ZFC+CH. [9.14]

Therefore,

If ZF is consistent, then $ZFC \nvdash \neg CH$.

Thus, it is futile to try proving in ZFC that the Continuum Hypothesis is false. Also, one could now assume the Continuum Hypothesis to prove theorems without worrying that this assumption might be the cause of any inconsistency.

Gödel's proofs of [9.12] and [9.14] are difficult and we can give only the barest sketch of what he did. First, he built up a hierarchy of sets, step by step. These are called the *"constructible"* sets. Every model of ZF has the empty set ∅ in its universe (shown in Appendix B), and at the first step, the only constructible set is ∅. On each subsequent step, the new constructible sets are those that can be defined by a statement in $\mathcal{L}_{ZF}$ involving only the constructible sets that were defined in previous steps. To describe them as "constructible" is appropriate because each set is explicitly defined (or constructed) by a statement in $\mathcal{L}_{ZF}$. The "number" of steps in this process goes well beyond all natural numbers, but to be more specific about this would take us too far afield. (See the reference to ordinal numbers in Appendix B.) The *Axiom of Constructibility* says that every set is constructible. This assertion can be expressed by a sentence in $\mathcal{L}_{ZF}$, and we abbreviate it as "*AxCon*". Although it is called an "axiom", very few believe it is true in the real world of sets because it is too disrespectful of Need 1, the need to allow as many objects as possible to be sets. Gödel then added *AxCon* to the axioms of ZF, producing a new theory that we call "ZF+AxCon".

Gödel then proved that

    If ZF is consistent, so is ZF+AxCon.   **[9.15]**

We can only roughly indicate why [9.15] is true. Suppose ZF is consistent, and $\mathfrak{M}$ is a model of ZF. An axiom of ZF (the Subset Axiom Schema) says that any collection that can be defined from an existing set (using some condition expressible in $\mathcal{L}_{ZF}$) exists as a set. In other words, any constructible set exists in any model. Then from the universe of $\mathfrak{M}$, we discard all the sets that are not constructible, giving us a structure $\mathfrak{L}$ in which the universe L consists entirely of constructible sets. It turns out that L contains every set that ZF demands. For example, the union of two sets can be defined by a sentence in $\mathcal{L}_{ZF}$, so the union of two constructible sets is constructible, and is therefore a member of L. By meeting the demands of ZF, $\mathfrak{L}$ is a model of ZF. Moreover, *AxCon* is true in $\mathfrak{L}$, so $\mathfrak{L}$ is a model of ZF+AxCon. Therefore, ZF+AxCon is consistent (assuming that ZF is consistent).

Gödel also proved that

    AC is true in $\mathfrak{L}$,

and   *CH* is true in $\mathfrak{L}$.

We can only roughly indicate why *AC* is true in $\mathfrak{L}$. L (the universe of $\mathfrak{L}$) is very well-organized. Each set in L was added at a certain step, and each set added at that step corresponds to a statement in $\mathcal{L}_{ZF}$, the one that defines it. Suppose $\mathbb{X}$ is a set of non-empty disjoint sets. Each set in $\mathbb{X}$ along with $\mathbb{X}$ itself are members of L (because in $\mathfrak{L}$, nothing else exists). $\mathbb{X}$ and its members being effectively sorted, the method of choosing the members of a choice set for $\mathbb{X}$ can be specifically defined. This produces a choice set that is itself constructible, and hence a member of L. The members of L are precisely the sets that exist in $\mathfrak{L}$. Thus, *AC* is true in $\mathfrak{L}$. (Recall that it was the vagueness of "choose a member from each set" that caused some to reject the Axiom of Choice, but in choosing members from L, there is no such vagueness.)

We will simply accept that *CH* is true in $\mathfrak{L}$ because Gödel's proof is too intricate for us to say anything meaningful about it. Here is one of the complicating factors. Suppose $\mathfrak{M}$ is a model of ZF, and suppose the assertion that

$\qquad$ $\mathbb{X}$ and $\mathbb{Y}$ are equipotent $\hfill$ [9.16]

is true in $\mathfrak{M}$. That means there exists in the universe of $\mathfrak{M}$ a one-to-one pairing between their members. (In a model of a theory, something "exists" means it is in the universe of the model.) But when we cut $\mathfrak{M}$ back to a smaller model $\mathfrak{L}$ with universe L, it is possible that $\mathbb{X}$ and $\mathbb{Y}$ are in L while all the pairing functions between them are not. In that case, there is no pairing function in L between $\mathbb{X}$ and $\mathbb{Y}$, and the sentence asserting the existence of a pairing function is false in $\mathfrak{L}$. Therefore, [9.16] could be true in $\mathfrak{M}$ but false in $\mathfrak{L}$. We intuitively think that two sets are either equipotent or not. But whether or not the sentence asserting they are equipotent is true in a model depends on the sets themselves *and* on the model; equipotency is not always an absolute fact.

We are almost there. $\mathfrak{L}$ is a model of ZF+AxCon in which *AC* and *CH* are both true. So $\mathfrak{L}$ is a model of both ZFC and ZFC+CH, showing that both these theories are consistent. The only assumption we made to reach this conclusion is that ZF is consistent. We have now "shown" how [9.12] and [9.14] were proven.

Although we usually use double-lined letters like "$\mathbb{L}$" to denote sets, we used "L" to denote the universe of $\mathfrak{L}$ because L is not a set; L is not a member of itself, and therefore not in the universe of $\mathfrak{L}$. If it were, then L would be "the set of all sets" in the model $\mathfrak{L}$, and that would lead to a contradiction.

9.10 Formal Undecidability of *AC* and *CH* in ZF

Gödel's theorems of 1938 showed that

$\qquad$ If ZF is consistent, then ZF $\nvdash \neg AC$

and $\quad$ If ZF is consistent, then ZFC $\nvdash \neg CH$.

In 1963, Paul Cohen proved the "other half" of what Gödel had proved. Cohen showed that

$\qquad$ If ZF is consistent, then ZF $\nvdash AC$

and $\quad$ If ZF is consistent, then ZFC $\nvdash CH$.

The theorems of Gödel and Cohen show that if ZF is consistent, then *AC* is formally undecidable in ZF, and *CH* is formally undecidable in ZFC (and therefore in ZF as well). In other words, ZF and ZFC are incomplete if they are consistent. This was known back in 1931 when the First Incompleteness Theorem had shown that there is a formally undecidable sentence in $\mathcal{L}_{ZF}$ (assuming set theory is consistent). But this sentence was deliberately constructed to be formally undecidable, and what it asserted had interest only when interpreted through an artificial coding system. Thus, it had no "natural" mathematical meaning. With AC and CH, we have mathematically interesting sentences that are formally undecidable in set theory (if set theory is consistent). So Hilbert's challenge to settle the Continuum Hypothesis one way or another remained unanswered. Soon, matters were made even more unsettled by further discoveries of interesting propositions that are independent of set theory (i.e. formally undecidable in ZF or ZFC).

Two schools of thought arose among mathematicians regarding propositions independent of set theory. Many believed that there was no point in pursuing the question of whether or not the Continuum Hypothesis (for example) is "true in the real world of sets". These mathematicians took the position that the axioms of ZF captured all that human intuition can tell us about sets. Thus, a sentence like CH that is independent of set theory is also independent of our intuition. We may assume CH or $\neg CH$ as an axiom whenever we like, and prove the consequences, without committing ourselves regarding the "truth" of our assumption. We might call this "agnostic mathematics".

The other school (which included Gödel) believed that new axioms could be found to strengthen ZF, axioms that "should be true about sets" according to our intuition. After set theory is strengthened by additional axioms, CH may no longer be formally undecidable, and Hilbert's problem would be resolved.

Gödel himself believed that the Continuum Hypothesis is false in the real world of sets, even though he had proved that it is consistent with ZFC. He began examining "large cardinal" axioms that declared the existence of enormous sets, in the hope of proving $\neg CH$. (See Appendix C for an explanation of the term "cardinal".) Once again, an idea pioneered by Gödel turned out to be very fruitful, and large cardinal axioms led to the resolution of several statements that were independent of set theory. But this approach has not resolved the problem of the Continuum Hypothesis.

CHAPTER 10
COSMOLOGY

10.1 A Very Brief History of Time

Gödel's foray into cosmology grew out of his interest in Immanuel Kant's philosophy of time. Kant maintained that the passage of time was not an objective reality "in itself", but a construct of the human mind. When Kant was putting forth his ideas in the 18th century, scientists believed that time was a universal reality, measured by a celestial clock that ticked at the same rate everywhere. In their view, history unfolds in a sequence of instants of time. At each instant, the universe is in a well-defined state, a universal "now", and all observers agree on what is happening anywhere "right now".

In 1905, Einstein's Special Theory of Relativity forever altered this traditional concept of time. According to Special Relativity, two different observers could observe two different histories of the universe, and both have an equal claim to being correct. One observer might see two events happening at the same instant, so for this observer, they occurred in one single "state of the universe". But to another observer, one of the events occurred before the other, so they occurred in two different "states of the universe", or on two pages of history. The fact that both are correct means that time is no longer absolute; there is no universal "now". Furthermore, clocks tick at different rates depending on how they move. This is most dramatically illustrated by the Twin Paradox; two twins embark on different journeys, but when they re-unite, they are no longer the same age.

In 1915, Einstein's General Theory of Relativity introduced new complications. In this theory, mass and energy distort the fabric of space-time, causing the flow of time to be altered not just by motion, but by the proximity of matter. Close to a massive object, time slows down. The two theories of Relativity revolutionised the traditional idea of time, but Gödel showed that the revolution might be deeper than anyone suspected.

10.2 The Gödel Universe

To Gödel, the Theory of Relativity provided evidence in favour of Kant's position on the reality of the passage of time. But Gödel being Gödel, he was not merely going to write an essay supporting the Kantian philosophy of time. Instead, in what is arguably his most spectacular achievement, he showed in

1949 that it is theoretically possible for a universe to exist where the familiar "past-to-present-to-future" progression of time does not hold. "Possible" means that the General Theory of Relativity (the verified and accepted theory of cosmology) does not forbid the existence of this strange universe since it exactly obeys all the laws of the theory. Today, it is known as the "Gödel Universe". The traditional notion of time had already been upset twice by Relativity, but time in the Gödel Universe was radically different. When past and future are inseparably mixed, time travel is possible. In the Gödel Universe, it is possible for a traveller to make a round-trip journey, and arrive back at the starting point only to find a younger version of himself about to begin the journey that had just ended. Did the journey begin before or after it ended? (When time travel is possible, words like "before" and "after" should be used with care.)

Before (careful!) we look deeper into the Gödel Universe, we should examine what Gödel was aiming to do. A scientific theory becomes more credible when it does either of the following:

1. It makes predictions that are verified by subsequent experiments. For example, General Relativity predicted that gravity bends light, which was famously confirmed in 1919.
2. It unifies apparently diverse phenomena into a common framework. For example, Newton's Law of Universal Gravitation is a single equation that accurately describes how apples fall from trees and how the moon orbits the Earth, making both phenomena better understood.

Gödel's work did neither of these. It did not predict that somewhere in the universe, past and future are mixed; its significance did not depend on having a prediction confirmed by experimental evidence. Nor did it make any known phenomenon more understandable (quite the opposite in fact). It was not even a new theory because it used the equations of General Theory of Relativity, a theory from 1915. So what did Gödel do, and what does it prove?

Gödel constructed a model. He considered a universe consisting of a rotating cloud of dust, and showed that under certain conditions, time travel is possible in this universe. He showed it by imposing the laws of General Relativity upon his dust storm, meaning that he solved Einstein's equations relating the curvature of space-time to the distribution of mass and energy within his universe. This implies that in our present universe, time behaves the way it does (moving in a one-way flow from past to present to future) because of the way that our universe happens to be configured. Time does not necessarily have to behave like this. If our universe happened to be configured like Gödel's dust storm, time would be very different. The Gödel Universe is something Kant never imagined, but the mere possibility of time travel strengthens his position that the passage of time lacks an independent objective reality.

10.3 Closed Time-like Curves

Many sources (in print and online) refer to "closed time-like curves" to explain time travel. We now examine what this phrase means. By definition, a curve is a continuous one-dimensional geometric object in some space. For example, the path traced by a butterfly in flight is a curve in our 3-dimensional physical space. The path traced by a moving object is called its "trajectory". A butterfly's trajectory is difficult to represent on paper, so we consider a simpler example. Suppose a man is walking East along a straight road. He starts at A, briefly stops at B, and then walks back towards the West, ending his walk at C.

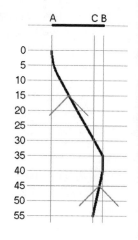

His trajectory is a straight line that folds back on itself as he retraces his steps. It is shown on the right as a heavy horizontal line, marked with points A, B, and C. The trajectory shows the route he took, but does not show when he reached which point along the way.

When a trajectory is displayed in a space with one extra dimension to show time, it is called a "world-line" in a space-time diagram. A world-line shows when the object was at each point of its trajectory. The world-line of the walking man is shown on the right. It is a curve in a 2-dimensional space, one dimension for space, and one for time. East is to the Right, and Future is Down. The space-time diagram shows that the man took 35 sec to go from A to B. There he stood for 5 sec, and then took another 15 sec to reach C.

The sides of the two "tents" in the diagram are the world-lines of four flashes of light. 15 sec after leaving A, the walker struck a match, and the light of the spark moved to the East and West as two flashes of light. 30 sec later, he did it again, creating another tent on his world-line.

On a space-time diagram, time is represented as a spatial dimension. The natural scale on the time axis is: t units of time equal ct units of distance, where c is the speed of light. (For a similar reason, m units of mass equal mc^2 units of energy.) In a time duration of t, a flash of light moves a distance of ct, so the distance travelled equals the travelling time. Therefore, a flash of light travels through equal "amounts" of space and time as it moves. So between any two points on the world-line of a flash of light, the difference in time is equal to the spatial separation. (We have graphically suggested this by sloping the world-lines of flashes of light at 45°.) An immutable law of Physics says that material objects always move slower than light, so they travel through less distance than time. Between any two points on the world-line of a material object, the difference in time is greater than the spatial separation. This is why world-lines of material objects are described as "time-like curves". For the same reason, they always remain "within the tent" as shown in the above diagram.

The simple situation shown in the above space-time diagram is consistent with Special Relativity. According to observers standing on the road, each horizontal line is a universal "now", and each vertical line is a history of one point along the road. But according to General Relativity, mass and energy distort the structure of space-time, causing it to curve (which also causes gravity). Over small regions of space and time, the effect of curvature is slight and the laws of Special Relativity apply. But over large regions, East cannot be represented by a fixed direction, and neither can Future. As an analogy, two people can be anywhere in Rome and if both point to the East, they are pointing in the same direction. But someone in Honolulu who points East is pointing in a different direction. Over a large distance, the curvature of the Earth's surface is revealed by this difference.

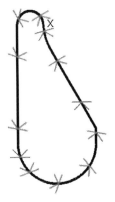

In the space-time diagram on the right, the man takes a much longer journey in which the curvature of space-time makes itself apparent. The surface of the Earth is 2-dimensional, and we can readily see the effect of its curvature. Space-time is a 4-dimensional structure, and we cannot visualize its curvature. Unless we do a mathematical analysis, the best we can do is think in terms of analogies. When space-time is curved, then East may not be in the same direction when you move to a new location. (Think of Rome and Honolulu.) As a result, long horizontal lines on a space-time diagram do not all align East-West. But in any small region of space-time, some fixed direction points East, and another fixed direction points to the Future. (Just don't leave Rome.) So the long horizontal lines in the first diagram are replaced by short ones that have different slopes. Just as East is not always to the Right, Future is not always Down. To determine the direction of Future, we have our man strike a match. In general, the resulting tent will lean to one side, and the centre-line of the tent (which is not shown) points to the Future in that small region of space-time. The immutable law of Physics still dictates that a world-line must remain within each little tent at every point. The point X marks the starting point of the journey, both in space and in time. As the journey unfolds, the curvature of space-time makes the tents tilt more and more until they are upside-down relative to the tent at X. To obey the laws of Physics and remain within the tents, the world-line is forced to go in the Up direction. This does not mean the man is getting younger because the Future is not Down and the Past is not Up. Since his speed is always less than the speed of light, he is always moving more through time than through space. In other words, his world-line is still a time-like curve. But the curvature of space-time makes it possible for his world-line to close back on itself in a loop and become a closed time-like curve. Gödel showed that this can happen in the Gödel Universe. (Our diagram only illustrates how "tent-tilting" caused by space-time curvature can lead to closed time-like curves. There are no such curves in a 2-dimensional space-time.)

When the man arrives back at X, he is at the same *place and time* that he was when he began his journey, so he sees his younger self about to start off. Gödel had shown that travelling back in time is consistent with the laws of Physics. This creates difficulties. Suppose the younger man is so intrigued by his older self that he decides to cancel his journey and just talk to him (or to himself?). (It's not every day that you meet yourself at a different age.) Then the journey never happens, and the older man would not arrive at X again, so they could not possibly have the conversation that they are already having. Conundrums like this have not been satisfactorily resolved.

10.4 More Thoughts on the Gödel Universe

The first exact solution of Einstein's equations of General Relativity was found by Karl Schwarzchild in 1915, the same year in which the equations were published. He considered the simple but important example of a universe where a single massive non-rotating sphere is situated in empty space. Einstein's equations are complex, and what Schwarzchild achieved was no mean feat. But what Gödel accomplished may be even more impressive. Gödel began with a feature that he suspected the universe could have, namely, closed time-like curves. Then, his intuitive understanding of Einstein's equations led him to configure his universe so that it possessed this feature. Finally, he proved his intuition was correct by solving the equations for this complex configuration, and showing that closed time-like curves do indeed exist. The preeminent physicist Paul Dirac wrote, "*I consider that I understand an equation when I can predict the properties of its solutions, without actually solving it.*" Gödel's ability to "design" a universe shows that his understanding of the equations of General Relativity met Dirac's lofty criterion. Between 1915 and 1949, neither Einstein nor a generation of physicists had thought to investigate a universe consisting of a rotating cloud of dust. Only Gödel saw that the solution to Einstein's equations for such a universe might have very unusual properties.

The discoveries of the Gödel Universe and the Incompleteness Theorems both began as bold visions that had eluded others, and both required meticulous work to turn the initial vision into a concrete result. Dawson (p. 184) points out more similarities. Both were totally unexpected, and not immediately accepted or appreciated. Both had profound implications on a theoretical and philosophical level, but almost no impact on the work that mathematicians and physicists did.

The Gödel Universe may be described as a "consistency" result with a mathematical flavour. Compare the two statements below:

> The Continuum Hypothesis is consistent with the axioms of set theory. It is true in the universe of constructible sets.

> The existence of closed time-like curves is consistent with the equations of General Relativity. Such curves exist in the Gödel Universe.

Scientists are usually more interested in the real world than in "possible" worlds that are merely consistent with theory. But cosmologists have shown more willingness than other scientists to consider speculative ideas such as parallel universes, wormholes, and time travel. Within this milieu, the Gödel Universe is right at home. Where Gödel's work will eventually stand as a piece of scientific research may still be uncertain. But it is without doubt a towering achievement requiring both imaginative vision and technical skill.

CHAPTER 11
ODDS AND ENDS

11.1 Quantity and Quality

Gödel's publications fill the first two volumes of *Kurt Gödel: Collected Works* (Oxford University Press). The next three volumes contain his unpublished manuscripts, lecture notes, essays, and correspondence. Two volumes of published papers may seem like a lot, yet Gödel published less than he could have. Being cautious by nature, he submitted papers only after much consideration and careful revision. As a result, his published works are unfailingly significant and influential. Very few mathematicians fill two volumes with work of the highest quality. Briefly, we mention below some of Gödel's other findings. These tend to be overlooked amidst the embarrassment of riches that we have with Gödel's output. Each of them might have been considered a major achievement if discovered by some other mathematician.

11.2 Computer Science

Some of Gödel's work may be categorised today as computer science. In 1936, he published *"On the Length of Proofs"*, in which he states a theorem concerning the minimum lengths of proofs in arithmetic. This was the first example of a "speed-up theorem" which today is an ongoing topic of research for computer scientists.

The central problem in computational complexity theory today is known as the "$P = NP$" question. Roughly speaking, it asks if a problem for which a solution can be quickly verified is necessarily one for which a solution can be quickly found. (We will not go into the precise definition of "quickly".) For example, finding the prime factors of a large number is not a quick process, but verifying the factors (if they are given) is quick. The question was formulated and clarified by Stephen Cook (and others) in 1971, but remains unsolved. Yet fifteen years earlier, in 1956, Gödel had raised a very similar question in a letter to von Neumann. Unfortunately, von Neumann was in the late stages of a terminal illness, and did not respond. That letter has now been romanticised as "Gödel's Lost Letter". It provides yet more evidence of Gödel's unrivalled ability to identify questions of fundamental importance. But on this question, he made no progress.

11.3 The Church-Turing Thesis

In trying to capture the informal notion of what is computable by *any* finitary process, Alonzo Church devised a system called the "λ-calculus" in 1936. (λ is the Greek letter lambda.) In the same year, with the same goal, Alan Turing defined a theoretical calculating device known as a "Turing machine". The functions that could be calculated by these systems were called "λ-computable" and "Turing-computable". Although the two systems were very different, the λ-computable functions turned out to be exactly the Turing-computable ones. The Church-Turing Thesis of 1936 declares that this class of functions is precisely the class of computable functions. Any function that can ever be computed by humans or by computers is λ-computable and Turing-computable. The Church-Turing Thesis identifies the vague intuitive notion of a "computable function" with the precise mathematically defined systems of the λ-calculus and Turing machines. It is called a "thesis" rather than a "theorem" because it can never be proved. However, it can be disproved; all it takes is for someone to find a function that is computable but not λ- or Turing-computable.

The Church-Turing Thesis has enormous significance in theoretical computer science. What can be more important than characterising exactly what is "computable" by any device, now and forever? In practice, the Thesis provides a means of proving negative results in what is decidable. For a decidable question, we only have to exhibit an algorithm (e.g. a computer program) that settles all instances of the question. But if a question is undecidable, and all we have is the vague intuitive notion of "computable", then there is no way to prove this negative result. All we can say is, "We have tried for years and failed, so the question is probably undecidable", which is hardly satisfactory. But with the Church-Turing Thesis, we have something well-defined that we can work with. If we can show that no Turing machine can decide a certain question, then it is undecidable according to the Thesis.

For example, consider the problem of determining if any given sentence in $\mathcal{L}_{PA}$ is provable in PA or not. The problem is *decidable* if there a routine procedure that can be applied to any sentence, and that always gives the correct answer "Provable" or "Not provable". To put it more precisely, consider the function *Prv* that takes a single number n as input, and outputs 1 if n is the Gödel number of a provable sentence in $\mathcal{L}_{PA}$, or outputs 0 if n is not. Is there a Turing machine that computes *Prv*? If there is, then the question of any sentence being provable or not in PA is decidable. Number theory would be a dead subject because anyone can tell if any sentence of $\mathcal{L}_{PA}$ is provable or not by simply giving its Gödel number to this Turing machine and waiting for the correct answer to emerge. Fortunately, the function *Prv* has been shown to be not Turing-computable. According to the Church-Turing Thesis, there is no effective procedure to determine if any given sentence in $\mathcal{L}_{PA}$ is provable or not. PA is an *undecidable theory*, and there will always be work for number theorists.

There are good reasons to believe the Church-Turing Thesis is correct. One reason is that several mathematicians have come up with their own definitions of what they believe is the class of computable functions, and they all turn out to be the same as the functions defined by Church or Turing. One such class, called the "general recursive functions", was defined by Gödel in 1934. He too was thinking about the nature of effective computability, and he found the same answer that Church and Turing found two years before they did. Many years later, there was a suggestion that he had formulated the Church-Turing Thesis before Church and Turing. But he denied this, saying that in 1934, he was not convinced that his general recursive functions were the correct characterisation of computability. The Thesis and the functions defined by it formed the basis of a major area of research at the intersection of logic and computer science, known as "recursion theory".

11.4 Intuitionism

Intuitionism was part of the constructivist response (see Section 2.3) to the crisis of confidence suffered by mathematicians around 1900. Logic under intuitionism was stricter than "normal" logic, and the more restrictive logic was designed to place Mathematics on a firmer footing, and avoid inconsistencies. Gödel was not an intuitionist, but he made significant contributions to intuitionist Mathematics. We mention one of his results in this field just to illustrate his versatility and the breadth of his interests.

When the underlying logic is intuitionistic logic, arithmetic is known as "Heyting Arithmetic", or "HA". The language of HA is the same as that of PA (first-order Peano Arithmetic), and the two theories have the same non-logical axioms (P1 to P7). But the logical axioms of HA are those of intuitionistic logic. For example, for *any* statement α, $\alpha \vee \neg \alpha$ is an axiom of PA but not HA. (This is the Law of the Excluded Middle which is not accepted by intuitionists.) Being based on a stricter logic, HA has less proving power than PA. Therefore, it is possible that HA is consistent while PA is not. But in 1933, Gödel proved that

if HA is consistent, so is PA. [11.1]

Despite using a safer logic, HA is not more immune from inconsistency than PA is.

In 1958, Gödel published a relative consistency proof for HA. Combined with [11.1], this was a proof of the consistency of PA. According to the Second Incompleteness Theorem, this proof cannot be carried out in PA. Indeed, the proof has unavoidable non-finitary steps. But it differs substantially from both Gentzen's proof and the consistency proof in ZFC. So now we have proofs based on three different systems that PA is consistent. Gödel's paper of 1958 also opened up important new avenues of research for constructivists.

11.5 Gödel's Loophole

Gödel's Loophole is not one of his major discoveries, but it may turn out to be his most consequential! Gödel came to the United States in 1933, escaping the unstable situation in Austria. In 1947, he applied for U.S. citizenship, a process that required a test be taken during an interview. Most people would regard such a test as almost a formality, but not Gödel. He studied in depth both the U.S. Constitution and the state laws of New Jersey where he lived. While doing so, he found an inconsistency in the Constitution, and told his friend Oskar Morgenstern (an economist at the Institute for Advanced Study) what he had discovered. Gödel believed that by exploiting this loophole, an elected leader could assume dictatorial powers. (Is this another example of his ability to identify significant problems that others had failed to notice?) He had already seen what dictators in Europe did in the 1930s after seizing power. Could it happen in the U.S.? When Morgenstern learned that Gödel intended to raise this question during the citizenship interview, he and Einstein both thought that doing so might jeopardize his application. So they attended the interview with Gödel, hoping to prevent him from talking about dictators seizing power in America. Incredibly, the presiding judge asked Gödel that very question. He eagerly launched into his well-researched reply before the judge wisely changed the subject. His application for citizenship was approved. The inconsistency he found in the Constitution is known today as "Gödel's Loophole", but nobody knows what it is. Researchers today are still examining the Constitution looking for the Loophole.

11.6 Final Thought

With an unerring ability to identify problems of mathematical and philosophical importance, Gödel made discoveries with deep implications for the foundations of Mathematics. Some of them were based on original ideas of dazzling ingenuity. He also displayed impressive technical skill and inventiveness in turning his initial inspirations into mathematical proofs. His trail-blazing discoveries and innovative methods opened up several branches within Mathematical Logic, uncovering fertile new fields of research. But he would usually leave it to others to reap the harvest while he moved on to his next big idea. He was the most creative, brilliant, and influential logician ever seen, a genius for the ages.

APPENDIX A

THE FIRST-ORDER INDUCTION AXIOM SCHEMA

When we assign a particular statement to be $\varphi(x)$, then

$$[\varphi(0) \ \& \ \forall k \ (\varphi(k) \to \varphi(s(k))) \ ] \ \to \ \forall x \ \varphi(x) \qquad [5.9]$$

becomes a P7-axiom that says something about $\{x : \varphi(x)\}$, the set of objects in the universe that make $\varphi(x)$ true. We analyse what [5.9] says:

$\varphi(0)$ says that **0** makes $\varphi(x)$ true, or $\{x : \varphi(x)\}$ includes 0.

$\forall k \ (\varphi(k) \to \varphi(s(k)))$ says that for all k, if k makes $\varphi(x)$ true, so does $s(k)$, or $\{x : \varphi(x)\}$ includes the successor of anything in $\{x : \varphi(x)\}$.

$\forall x \ \varphi(x)$ says everything in the universe makes $\varphi(x)$ true, or $\{x : \varphi(x)\}$ is the whole universe.

Putting it all back together, [A.1] below expresses informally what [5.9] says.

If [$\{x : \varphi(x)\}$ includes 0, and includes the successor of anything in $\{x : \varphi(x)\}$], then $\{x : \varphi(x)\}$ is the whole universe. **[A.1]**

When expressed informally, any P7-axiom is [A.1] with a particular statement chosen for $\varphi(x)$. Any subset of ℕ that includes 0, and includes the successor of anything in the subset is all of ℕ, the whole universe of 𝔑. Therefore, every P7-axiom is true in 𝔑, so 𝔑 is a model of PA.

We now find a P7-axiom that is false in $\mathfrak{Q}^+$. Suppose $\varphi(x)$ is [5.10] below:

$$\neg \ x + x = s(0) \qquad [5.10]$$

With [5.10] as $\varphi(x)$, [5.9] becomes the P7-axiom [5.11] (shown in Section 5.3). [5.10] says that $x + x$ is not 1. So in any structure, $\{x : \varphi(x)\}$ is the set of everything in the universe except 0.5. We call this set "NotHalf". With [5.10] as $\varphi(x)$, [A.1] becomes [A.2] below. [A.2] informally expresses [5.11], the P7-axiom corresponding to [5.10].

If [NotHalf includes 0, and includes the successor of anything in NotHalf], then NotHalf is the whole universe. **[A.2]**

In the structure $\mathfrak{Q}^+$, NotHalf includes 0, and includes the successor of anything in NotHalf. But NotHalf is not the whole universe because the universe of $\mathfrak{Q}^+$ includes 0.5, and NotHalf does not. Therefore, [A.2] is false in $\mathfrak{Q}^+$, and so is the P7-axiom [5.11]. Hence, $\mathfrak{Q}^+$ is not a model of PA.

Note that in 𝔑, NotHalf is ℕ (the universe of 𝔑). So [A.2] is true in 𝔑.

APPENDIX B

ZERMELO-FRAENKEL SET THEORY

To acquire a feeling for ZF, we look at some of its axioms, and see how they immunise set theory against Cantor's Paradox and Russell's Paradox. We will also see how every model of ZF includes a copy of the natural numbers. First, a convenient abbreviation: for any statements α and β,

"$\alpha \leftrightarrow \beta$" abbreviates "$\alpha \to \beta$ & $\beta \to \alpha$" (or informally, α if and only if β),

which means that α and β are either both true or both false.

The language of ZF (denoted by "$\mathcal{L}_{ZF}$") has the Equality symbol $=$ and one non-logical predicate symbol ε. The predicate ε is intended to mean "__ is a member of __", and the axioms of ZF are designed to capture the notion of a set by specifying conditions that ε satisfies.

The *Extensionality Axiom* captures the idea of when two sets are the same:

$\forall x \, \forall y \, (\, \forall z \, (\, z \, \varepsilon \, x \leftrightarrow z \, \varepsilon \, y \,) \to x = y \,)$

This says that if two sets have exactly the same members, then they are the same set. This is an essential characteristic of what we mean by "set".

If this were the *only* axiom of ZF, we could form a model of ZF by taking the natural numbers as the universe and choosing the "__ is a factor of __" predicate for the role of ε. But we cannot choose the "__ is a prime factor of __" predicate because, for example, 6 and 12 have the same prime factors, which means that two different sets have exactly the same members.

When working in set theory informally, we often start with a set, and form a new set by naming a condition; the new set consists of all the members of the starting set that satisfy the condition. For example, we may start with the set of natural numbers, and apply the condition "n is even". This gives us the set of even natural numbers. There should be an axiom of ZF that declares the existence of a new set based on any starting set and any expressible condition. We want the axiom to say something like this:

$\forall x \, \forall\text{statement } \theta(w) \, \exists y \, \forall z \, (\, z \, \varepsilon \, y \leftrightarrow (\, z \, \varepsilon \, x \, \& \, \theta(z) \,) \,)$ [B.1]

In [B.1], x is any starting set, and $\theta(w)$ is any condition. [B.1] declares that there is a set y that consists of all the members of x that satisfy condition $\theta(w)$. But [B.1] is not a first-order sentence because of "$\forall\text{statement } \theta(w)$", so it is not a sentence in $\mathcal{L}_{ZF}$. It is not acceptable as an axiom of ZF, but don't despair. We have already seen how to get around a difficulty like this: use an axiom schema. So we have the *Subset Axiom Schema*:

> For every statement $\theta(w)$ in $\mathcal{L}_{ZF}$ with one free variable, the following sentence is an axiom of ZF: $\forall x \exists y \forall z (z \varepsilon y \leftrightarrow (z \varepsilon x \,\&\, \theta(z)))$ **[B.2]**

[B.2] is an axiom schema (like Axiom P7 in PA). It generates an infinite number of axioms of ZF, one for each statement $\theta(w)$ in $\mathcal{L}_{ZF}$.

Both Cantor's Paradox and Russell's Paradox emerge when we work with the "set of all sets". The Subset Axiom Schema can show that there is no such set in any model. Suppose $\mathfrak{M}$ is any model of ZF and $\mathbb{X}$ is any object in the universe of $\mathfrak{M}$. Let $\theta(w)$ be $\neg\, w \,\varepsilon\, w$. Then [B.3] below is true in $\mathfrak{M}$ because it is one of the axioms in the Subset Axiom Schema:

$$\forall x \exists y \forall z (z \varepsilon y \leftrightarrow (z \varepsilon x \,\&\, \neg\, z \varepsilon z)).$$ **[B.3]**

By definition of $\forall$, we may delete $\forall x$ and replace all remaining x by any object in the universe (e.g. $\mathbb{X}$) to give us a true sentence. So [B.4] below is true in $\mathfrak{M}$.

$$\exists y \forall z (z \varepsilon y \leftrightarrow (z \varepsilon \mathbb{X} \,\&\, \neg\, z \varepsilon z))$$ **[B.4]**

By definition of $\exists$, there is an object $\mathbb{Y}$ in the universe of $\mathfrak{M}$ such that [B.5] is true in $\mathfrak{M}$. (Unlike $\forall$-sentences, $\mathbb{Y}$ is not "any object"; [B.4] merely says that there is at least one object $\mathbb{Y}$ in the universe of $\mathfrak{M}$ that makes [B.5] true.)

$$\forall z (z \varepsilon \mathbb{Y} \leftrightarrow (z \varepsilon \mathbb{X} \,\&\, \neg\, z \varepsilon z))$$ **[B.5]**

Again by definition of $\forall$ (replacing every z by $\mathbb{Y}$), [B.6] is true in $\mathfrak{M}$.

$$\mathbb{Y} \varepsilon \mathbb{Y} \leftrightarrow (\mathbb{Y} \varepsilon \mathbb{X} \,\&\, \neg\, \mathbb{Y} \varepsilon \mathbb{Y})$$ **[B.6]**

Now we assume that $\mathbb{Y} \varepsilon \mathbb{Y}$ (which will lead to a contradiction). By definition of $\leftrightarrow$, $(\mathbb{Y} \varepsilon \mathbb{X} \,\&\, \neg\, \mathbb{Y} \varepsilon \mathbb{Y})$ is true. By definition of &, $\neg\, \mathbb{Y} \varepsilon \mathbb{Y}$ is true, which contradicts our assumption that $\mathbb{Y} \varepsilon \mathbb{Y}$ is true. Therefore, $\mathbb{Y} \varepsilon \mathbb{Y}$ is false, and [B.6] tells us that $(\mathbb{Y} \varepsilon \mathbb{X} \,\&\, \neg\, \mathbb{Y} \varepsilon \mathbb{Y})$ is also false. But since $\neg\, \mathbb{Y} \varepsilon \mathbb{Y}$ is true and $(\mathbb{Y} \varepsilon \mathbb{X} \,\&\, \neg\, \mathbb{Y} \varepsilon \mathbb{Y})$ is false, we conclude that $\mathbb{Y} \varepsilon \mathbb{X}$ must be false in $\mathfrak{M}$. Therefore, there is an object $\mathbb{Y}$ that is not a member of $\mathbb{X}$. But $\mathbb{X}$ was an arbitrarily chosen object, so we conclude that no object in the universe of $\mathfrak{M}$ contains everything in the universe of $\mathfrak{M}$. Since there is no "set of all sets" in $\mathfrak{M}$, there is no Cantor's Paradox or Russell's Paradox.

Suppose $\mathfrak{M}$ is a model of ZF and $\mathbb{X}$ is any object in the universe of $\mathfrak{M}$. Let $\theta(w)$ be $\neg\, w = w$. Then [B.6] is true in $\mathfrak{M}$ because of the Subset Axiom Schema:

$$\forall x \exists y \forall z (z \varepsilon y \leftrightarrow (z \varepsilon x \,\&\, \neg\, z = z))$$ **[B.6]**

Then as we saw above with [B.5], there is an object $\mathbb{Y}$ in the universe of $\mathfrak{M}$ such that [B.7] is true in $\mathfrak{M}$.

$$\forall z\,(\,z\,\varepsilon\,\mathbb{Y} \leftrightarrow (\,z\,\varepsilon\,\mathbb{X}\,\&\,\neg z = z\,)\,) \qquad \text{[B.7]}$$

But $(\,z\,\varepsilon\,\mathbb{X}\,\&\,\neg z = z\,)$ is false in $\mathfrak{M}$ for every z. Therefore,

$z\,\varepsilon\,\mathbb{Y}$ is false in $\mathfrak{M}$ for every z.

Therefore, $\mathbb{Y}$ has no members. In other words, every model of ZF has an object in its universe with no members, which is the empty set (denoted by "$\emptyset$").

Axioms are easier to understand when expressed in English rather than in $\mathcal{L}_{ZF}$. The *Pairs Axiom* says that if $\mathbb{X}$ and $\mathbb{Y}$ are in the universe, there is another object whose only members are $\mathbb{X}$ and $\mathbb{Y}$. We denote it as "$\{\mathbb{X},\mathbb{Y}\}$". If $\mathbb{X}$ and $\mathbb{Y}$ are the same object then $\{\mathbb{X},\mathbb{Y}\}$ has only one member, and is usually denoted as "$\{\mathbb{X}\}$" (or "$\{\mathbb{Y}\}$"). Therefore, if $\mathbb{X}$ is in the universe of a model of ZF, so is $\{\mathbb{X}\}$.

A simplified version of the *Union Axiom* says that if $\mathbb{X}$ and $\mathbb{Y}$ are in the universe, so is the union of $\mathbb{X}$ and $\mathbb{Y}$, which is denoted as "$\mathbb{X} \cup \mathbb{Y}$".

Now we repeatedly use the Pairs Axiom and the Union Axiom. Since $\emptyset$ is in the universe, so is $\{\emptyset\}$, so is $\{\{\emptyset\}\}$, so is $\{\emptyset\}\cup\{\{\emptyset\}\}$ (which is $\{\emptyset,\{\emptyset\}\}$), so is $\{\{\emptyset,\{\emptyset\}\}\}$, so is $\{\emptyset,\{\emptyset\}\}\cup\{\{\emptyset,\{\emptyset\}\}\}$ (which is $\{\emptyset,\{\emptyset\},\{\emptyset,\{\emptyset\}\}\}$), and so on. Hence, every object in the infinite list below is in the universe of every model of ZF, and they may be taken as representations of the natural numbers:

$\emptyset$, $\{\emptyset\}$, $\{\emptyset,\{\emptyset\}\}$, $\{\emptyset,\{\emptyset\},\{\emptyset,\{\emptyset\}\}\}$, $\{\emptyset,\{\emptyset\},\{\emptyset,\{\emptyset\}\},\{\emptyset,\{\emptyset\},\{\emptyset\,\{\emptyset\}\}\}\}$, etc.

0,   1,   2,   3,   4, etc.

Hence, $0 = \emptyset$, $1 = \{0\}$, $2 = \{0, 1\}$, $3 = \{0, 1, 2\}$,..., $n + 1 = \{0, 1, 2,..., n\}$. More concisely, $0 = \emptyset$, and for all natural numbers n,

$$n + 1 = n \cup \{n\} \qquad \text{[B.8]}$$

For example, $3 + 1 = 4 = \{0, 1, 2, 3\} = \{0, 1, 2\} \cup \{3\} = 3 \cup \{3\}$.

The *Axiom of Infinity* says that there is a set that contains $\emptyset$, and also contains $\mathbb{X} \cup \{\mathbb{X}\}$ whenever it contains $\mathbb{X}$. So every model of ZF has a set $\mathbb{I}$ in its universe that contains $\emptyset$ (which is 0), and contains $0 \cup \{0\}$ (which is 1), $1 \cup \{1\}$ (which is 2), $2 \cup \{2\}$ (which is 3), and so on. Thus, $\mathbb{I}$ contains every natural number. The Subset Axiom Schema with $\mathbb{I}$ as the starting set and a messy condition θ(w) shows that in the universe of every model of ZF, there is a set that contains *only* the natural numbers. This set is denoted by "ω". Thus,

$\omega = \{0, 1, 2, 3,...\}$.

We continue to use the Pairs Axiom and Union Axiom as we did before when we generated the natural numbers in any model of ZF. Since ω is in the universe of every model of ZF, so is $\{\omega\}$, and so is $\omega \cup \{\omega\}$. If we consider [B.8] to be the definition of $n + 1$ for *any* set n, then

$$\omega + 1 = \omega \cup \{\omega\} = \{0, 1, 2, 3, \ldots \omega\},$$

and $\omega + 2 = (\omega + 1) + 1 = (\omega + 1) \cup \{\omega + 1\} = \{0, 1, 2, 3, \ldots \omega, \omega + 1\}$.

Both these are objects in the universe of every model of ZF, as well as $\omega + 3$, $\omega + 4$, and so on. Other axioms of ZF ensure that this process continues indefinitely. These sets are known as the *"ordinal numbers"*. They are theoretically significant, and often useful in proofs. For example, the steps in Gödel's definition of the constructible sets are enumerated by the ordinal numbers. Listed below are a few more ordinal numbers.

$$\omega + \omega \text{ (or } \omega \cdot 2), \quad \omega \cdot 2 + 1, \quad \omega \cdot 3, \quad \omega \cdot \omega \text{ (or } \omega^2), \quad \omega^3, \quad \omega^\omega, \quad \omega^{\omega \cdot \omega}$$

The functions + and · informally mentioned above may be properly defined as functions on all ordinal numbers. It may be shown that their effect on finite ordinal numbers is consistent with addition and multiplication on the natural numbers. Therefore, we may construct a model of PA in any model of ZF. Take ω as the universe, and make the obvious choices for **0**, **s**, **+**, and **×**, the non-logical symbols of $\mathcal{L}_{PA}$. (The choice for **s** is given by [B.8].)

There is more than one way of representing the natural numbers in models of ZF. The way that was shown above is particularly useful because it extends naturally (through [B.8]) to infinite ordinal numbers. This definition of ordinal numbers in set theory is due to von Neumann.

In this Appendix, we have explicitly mentioned where we used various axioms of ZF since this may be the first time you have encountered ZF. In practice, a set theorist never bothers to mention, for example, that the Pairs Axiom is being used at a certain point in a proof. But since we are being pedantically careful, we should note that we were sloppy on a few occasions. Whenever we give a name to something we have described, we should check that there is one and only one thing fitting that description, For example, we should not give a name to "the largest natural number", or to "the natural number between 2 and 5". For the sets that we described and named (e.g. $\emptyset$, ω, $\{\mathbb{X}, \mathbb{Y}\}$, $\mathbb{X} \cup \mathbb{Y}$), various axioms of ZF ensure that at least one set fits each description. But we should have mentioned in each case that the Extensionality Axiom ensures uniqueness.

APPENDIX C
CARDINAL NUMBERS

The relationship of being equipotent generalizes to all sets the relationship of being the "same size" for finite sets. But we have not generalized the concept of "size" itself to cover all sets. The reason for delaying this generalization is the difficulty with proving [9.10] below:

For any sets $\mathbb{X}$ and $\mathbb{Y}$, either $\mathbb{X} \preccurlyeq \mathbb{Y}$ or $\mathbb{Y} \preccurlyeq \mathbb{X}$. **[9.10]**

[9.10] is equivalent to [C.1] below:

For any sets $\mathbb{X}$ and $\mathbb{Y}$, either $\mathbb{X} \prec \mathbb{Y}$, or $\mathbb{Y} \prec \mathbb{X}$, or $\mathbb{Y} \equiv \mathbb{X}$. **[C.1]**

Without the Axiom of Choice, [9.10] and [C.1] are both unprovable. If [C.1] were false, then for some sets $\mathbb{X}$ and $\mathbb{Y}$, all three alternatives in [C.1] are false. Then we would have two sets with different "sizes", yet neither is "smaller than" the other. This is not a satisfactory situation.

So now we assume the Axiom of Choice. Now [C.1] is provable and the unsatisfactory situation can no longer arise; our concept properly captures the notion of "size" for all sets. We use the word "*cardinality*" instead of "size" to indicate that we are referring to the size of a set according to the criteria of "$\equiv$" and "$\prec$". As we did for size of finite sets, the cardinality of any set $\mathbb{X}$ is symbolized by "$|\mathbb{X}|$". Thus, "$|\mathbb{N}|$" and "$|\mathbb{E}|$" denote the cardinality of $\mathbb{N}$ and $\mathbb{E}$ respectively. Since these two sets have equal cardinality, we have

$|\mathbb{N}| = |\mathbb{E}|$, which is another way of saying $\mathbb{N} \equiv \mathbb{E}$.

Also, for any set $\mathbb{X}$, $|\mathbb{X}| < |\mathcal{P}(\mathbb{X})|$, which is another way of saying $\mathbb{X} \prec \mathcal{P}(\mathbb{X})$.

For finite sets, we have ready-made symbols to express the cardinality of a set. For example,

$|\{5, 7, 8, 9\}| = 4$.

In expressing the cardinality of a set, "4" in the above equation represents a *cardinal number*, or simply a *cardinal*. So every natural number can serve a second purpose as a cardinal number.

We now need symbols to express the cardinality of infinite sets. For the smallest infinite set $\mathbb{N}$, Cantor chose "$\aleph_0$". This symbol is pronounced as "aleph-zero", $\aleph$ being the first letter of the Hebrew alphabet. Therefore,

$|\mathbb{N}| = \aleph_0$.

Since $|\mathbb{N}| = |\mathbb{E}|$, we also have

$|\mathbb{E}| = \aleph_0$.

But we have many more infinite sets. The sets listed below are all infinite, and each set is smaller than the next because every set is smaller than its power set:

$\mathcal{P}(\mathbb{N}), \quad \mathcal{P}(\mathcal{P}(\mathbb{N})), \quad \mathcal{P}(\mathcal{P}(\mathcal{P}(\mathbb{N}))), \quad \mathcal{P}(\mathcal{P}(\mathcal{P}(\mathcal{P}(\mathbb{N})))), \ldots \text{etc.}$ [C.2]

To express the cardinality of these sets, we use notation that expresses our knowledge of finite sets. When $\mathbb{X}$ is a finite set, then so is $\mathcal{P}(\mathbb{X})$, and the following equation is a fact relating two finite numbers:

$|\mathcal{P}(\mathbb{X})| = 2^{|\mathbb{X}|}$. [C.3]

For example, when $\mathbb{X}$ is a set with 3 members, $\mathcal{P}(\mathbb{X})$ has 2^3 (or 8) members.

We want to choose a symbol to express the cardinality of $\mathcal{P}(\mathbb{N})$. There are many possible choices, but perhaps the best choice is made using [C.3]. We simply declare that [C.3] is true for all sets, including infinite ones. Then we have

$|\mathcal{P}(\mathbb{N})| = 2^{|\mathbb{N}|}$, and therefore,

$|\mathcal{P}(\mathbb{N})| = 2^{\aleph_0}$. [C.4]

$2^{\aleph_0}$ is a cardinal number; it is the cardinality of $\mathcal{P}(\mathbb{N})$, and also of $\mathcal{P}(\mathbb{E})$ and $\mathbb{R}$.

Note that 2^3, for example, means the result of multiplying 2 by itself 3 times. But $2^{\aleph_0}$ does *not* mean "the result of multiplying 2 by itself $\aleph_0$ times" (which is a meaningless phrase).

When $\mathbb{X}$ is a finite set, [C.3] is an equation between two natural numbers that have already been defined. [C.3] is a statement with mathematical content.

When $\mathbb{X}$ is an infinite set, we use [C.3] to *define* what the expression "$2^{|\mathbb{X}|}$" means. [C.3] and [C.4] are definitions with no mathematical content.

(A similar procedure is used to define the meaning of "$2^{0.5}$" using the equation $2^n \times 2^m = 2^{n+m}$.)

Combining [C.3] and [C.4], we can now express the cardinality of every set in the list [C.2] (although the expressions soon become unwieldy with repeated superscripts).

If $\mathfrak{a}$ and $\mathfrak{b}$ are cardinals, we say that $\mathfrak{b}$ is an *immediate successor* of $\mathfrak{a}$ if $\mathfrak{a} < \mathfrak{b}$ and there is no cardinal $\mathfrak{c}$ such that $\mathfrak{a} < \mathfrak{c} < \mathfrak{b}$. Assuming the Axiom of Choice, every cardinal has a unique immediate successor.

For finite cardinals, we don't need the Axiom of Choice; the immediate successor of n is $n + 1$. We denote the immediate successor of $\aleph_0$ by "$\aleph_1$", and its immediate successor by "$\aleph_2$", and so forth. Therefore,

$$\aleph_0 < \aleph_1 < \aleph_2 < \aleph_3 < \aleph_4 < \ldots < \aleph_n < \ldots \text{etc}\ldots$$

and there are no cardinal numbers that can squeeze in between two adjacent terms in this chain of inequalities. In other words, for any natural number n,

There is no cardinal $\mathfrak{a}$ such that $\aleph_n < \mathfrak{a} < \aleph_{n+1}$. [C.5]

[C.5] just follows from the definition of $\aleph_{n+1}$.

The Continuum Hypothesis was stated as follows:

There is no set $\mathbb{X}$ such that $\mathbb{N} \prec \mathbb{X} \prec \mathcal{P}(\mathbb{N})$. [9.9]

Using the language of cardinals, [9.9] can be restated as

There is no cardinal $\mathfrak{a}$ such that $|\mathbb{N}| < \mathfrak{a} < |\mathcal{P}(\mathbb{N})|$,

or   There is no cardinal $\mathfrak{a}$ such that $\aleph_0 < \mathfrak{a} < 2^{\aleph_0}$,

or $2^{\aleph_0}$ is the smallest cardinal greater than $\aleph_0$.

But $\aleph_1$ is by definition the smallest cardinal greater than $\aleph_0$, and it is the only cardinal with this property. Therefore, the Continuum Hypothesis can be stated as follows:

$$2^{\aleph_0} = \aleph_1.$$

The Axiom of Choice plays an essential role in much of what we can say about cardinals. For example, with the Axiom of Choice, we know that

For any sets $\mathbb{X}$ and $\mathbb{Y}$, either $\mathbb{X} \prec \mathbb{Y}$ or $\mathbb{Y} \prec \mathbb{X}$ or $\mathbb{Y} \equiv \mathbb{X}$. [C.1]

Expressed in the language of cardinals, [C.1] says that

For any two cardinals $\mathfrak{a}$ and $\mathfrak{b}$, either $\mathfrak{a} < \mathfrak{b}$ or $\mathfrak{b} < \mathfrak{a}$ or $\mathfrak{b} = \mathfrak{a}$. [C.6]

With the Axiom of Choice, [C.6] is true, and any two immediate successors of a cardinal must be equal because neither can be less than the other. Without the Axiom of Choice, [C.6] might be false. Then we could have two different cardinals with both being an immediate successor of some cardinal (since each can be not less than the other without being equal).

Assuming the Axiom of Choice, we can make definitions of addition and multiplication for all cardinals. When applied to finite cardinals, these definitions are consistent with the usual addition and multiplication for natural numbers.

INDEX

&, 12
&-Elimination Rule, 12
=, 16
∀, 14
∀-Elimination Rule, 14
∃, 15
∃-Introduction Rule, 15
≡, 69
≼, 70
≺, 71
⊢, 19
⊨, 20
∨, 13
→, 13
→-Elimination Rule, 12
¬, 13
∅, 68, 96
ω, 96
ω-consistency, 48
ℵ, 98

Ackermann, Wilhelm, 25, 36
arithmetic, 4
   axiomatization of, 30
axiom, 8, 12
   logical, 13, 22
   non-logical, 16
Axiom of Choice, 6, 11, 29, 74, 77, 98
   consistency of, 78
   formal undecidability of, 80
Axiom of Constructibility, 78
axiom schema, 33

Barber's Paradox, 5
Bernays, Paul, 57
Bohr, Niels, 1
Bolyai, János, 3

Boole, George, 3
Brouwer, L.E.J., 7

Cantor, Georg, 4, 10
   power set theorem, 71
Cantor's Paradox, 5, 94, 95
cardinal number, 98
choice set, 75
Church, Alonzo, 89
Church-Turing Thesis, 89
closed time-like curve, 84
Cohen, Paul, 80
Compactness Theorem, 27, 34, 35
Completeness Theorem, 22, 25, 56
consistent theory, 21
constructible sets, 78, 97
constructivism, 6, 90
Continuum Hypothesis, 73, 77, 100
   consistency of, 78
   formal undecidability of, 80
Cook, Stephen, 88
cosmology, 82

Dawson, John W. Jr., vi, 2, 26, 87
Dedekind, Richard, 10, 30
diagonalization, 55, 73
Dirac, Paul, 87

Einstein, Albert, v, 1, 82, 91
equipotent, 68
existential quantifier, 15
expressible property, 43
expressing truth in arithmetic, 52

finitary, 10
First Incompleteness Theorem, 38
   scope of, 49

first-order
  logic, 16
  theory, 16
formal proof, 8
formalism, 7, 39
formally undecidable, 23
Fraenkel, Abraham, 11
free variable, 15
Frege, Gottlob, 4, 10

Galois, Évariste, 17
Gauss, Carl Friedrich, 3
general recursive function, 90
Gentzen, Gerhard, 62, 65, 66, 91
Gödel numbering, 42
Gödel Universe, 82
Gödel, Kurt, v - 100
Gödel's Loophole, 91
Gödel's Lost Letter, 88
group theory, 17

HA. *See* Heyting Arithmetic
Hahn, Hans, 26
Heyting Arithmetic, 90
  consistency of, 91
Heyting, Arend, 7, 90
Hilbert, David, 1, 4, 7, 9, 25, 57

incomplete theory, 37
Incompleteness Theorems, 37
inconsistent theory, 21, 22
Induction Axiom, 30, 33, 35
  first-order, 93
intuitionism, 7, 90

Kant, Immanuel, 82
Kronecker, Leopold, 7, 66
König, Dénes, 29
König's Lemma, 28

Law of the Excluded Middle, 7

Liar Paradox, 40
Lobachevsky, Nikolai, 3
logicism, 10, 50
$\mathcal{L}_{PA}$, 31
$\mathcal{L}_{ZF}$, 35, 76, 94

mathematical theory, 16
metamathematics, 7
model, 16
Model Existence Theorem, 27
model theory, 26
Morgenstern, Oskar, 91

$\mathbb{N}$, 15
$\mathfrak{N}$, 30
$\langle\mathfrak{N}\rangle$, 52
natural number, 15
negation-complete theory, 23
negation-incomplete theory, 23, 37
non-Euclidean geometry, 3
numeral, 32

ordinal numbers, 97

$P = NP$, 88
PA. *See* Peano Arithmetic
Peano Arithmetic, 33
  and set theory, 35
  non-standard model, 31, 34
  proving consistency, 60, 61, 65, 91
  second-order, 33, 35
  standard model, 31
Peano Axioms, 31
Peano, Giuseppe, 4, 30
Peirce, Charles, 4
Poincaré, Henri, 4
Post, Emil, 25
power set, 71
predicate, 13
Presburger Arithmetic, 51

INDEX

Presburger, Mojżesz, 36, 51
PrfPA, 44
Principia Mathematica, 10, 38, 50, 62
proof, 12
proof theory, 7
propositional connective, 13
propositional logic, 13, 25
proving consistency, 21

quantifier
   existential, 15
   universal, 14
Quine, W.V., 30

reasonable theory of arithmetic, 38, 50
refutable sentence, 24
Rogers, Hartley Jr., 41
Rosser, J. Barkley, 49, 50
rules of inference, 8, 12, 13, 19
Russell, Bertrand, v, 5, 10
Russell's Paradox, 5, 10, 67, 94, 95

Schwarzchild, Karl, 86
Second Incompleteness Theorem, 57
self-referential definition, 5
sentence, 16
set theory, 10, 67
   and Peano Arithmetic, 35
   axiomatic, 76, 94
   Zermelo-Fraenkel, 11, 38, 76, 94

Skolem Arithmetic, 51
Skolem, Thoralf, 26, 36, 51
Soundness, 20
speed-up theorem, 88
statement, 15
structure, 16
successor function, 31
symbol
   logical, 16
   non-logical, 16

Tarski, Alfred, 54
Tarski's Theorem, 53, 54
term, 46
theorem, 12, 19
theory, 16
   effective, 49
   undecidable, 90
time travel, 83
Turing, Alan, 89

universal quantifier, 14
universe, 14, 16

valid, 20
variable, 14
von Neumann, John, 1, 11, 36, 51, 57, 88, 97

Whitehead, Alfred North, 10

Zermelo, Ernst, 6, 11, 51
ZF, 11, 76
ZFC, 11, 35, 38, 77

Made in the USA
Middletown, DE
26 April 2022